WORKBOOK
TO ACCOMPANY
MAINTENANCE and RELIABILITY BEST PRACTICES
SECOND EDITION

Ramesh Gulati
Christopher Mears

ISBN 978-0-8311-3435-8

Industrial Press, Inc.
32 Haviland Street
South Norwalk, Connecticut 06854

Copyright © 2014 by Industrial Press, Inc. All rights reserved. With the exception of quoting brief passages, no part of this publication may be reproduced or transmitted in any form without written permission from the publisher.

2 3 4 5 6 7 8 9 10

TABLE OF CONTENTS

Chapter 1 Introducing Best Practices   1
Self-Assessment Questions 8
Comprehension Assessment Questions 9
Research Questions 11

Chapter 2 Culture and Leadership   13
Self-Assessment Questions 29
Comprehension Assessment Questions 30
Research Questions 33

Chapter 3 Understanding Maintenance   35
Self-Assessment Questions 55
Comprehension Assessment Questions 56
Research Questions 58

Chapter 4 Work Management: Planning and Scheduling   59
Self-Assessment Questions 76
Comprehension Assessment Questions 77
Research Questions 80

Chapter 5 Materials, Parts, and Inventory Management   81
Self-Assessment Questions 96
Comprehension Assessment Questions 97
Research Questions 99

Chapter 6 Measuring and Designing for Reliability and Maintainability   101
Self-Assessment Questions 123
Comprehension Assessment Questions 124
Research Questions 126

Chapter 7 Operator Driven Reliability   127
Self-Assessment Questions 138
Comprehension Assessment Questions 139
Research Questions 140

Chapter 8 Maintenance Optimization   141
Self-Assessment Questions 164
Comprehension Assessment Questions 165
Research Questions 167

Chapter 9   Managing Performance   **169**
Self-Assessment Questions 180
Comprehension Assessment Questions 181
Research Questions 182

Chapter 10   Workforce Management   **183**
Self-Assessment Questions 198
Comprehension Assessment Questions 199
Research Questions 200

Chapter 11   Maintenance Analysis and Improvement Tools   **203**
Self-Assessment Questions 219
Comprehension Assessment Questions 220
Research Questions 222

Chapter 12   Current Trends and Practices   **223**

Appendix   **241**
Self-Assessment Questions and Answers 241

Chapter 1
Introducing Best Practices

"I have not failed. I've just found 10,000 ways that won't work."

-Thomas Edison

Chapter 1: Introducing Best Practices

1.1 Introduction: What Is a Best Practice?
1.2 Key Terms and Definitions
1.3 What Do Best Practices Have to Do with Maintenance and Reliability?
1.4 Examples of Maintenance and Reliability Benchmarks
1.5 Basic Test on Maintenance and Reliability Knowledge
1.6 Summary
1.7 Self Assessment Questions
1.8 References and Suggested Readings

Chapter's Objective ... to understand:

- What are best practices and why do we care?
- What do best practices have to do with maintenance and reliability?
- Key Maintenance and Reliability (M&R) terms and benchmark examples

You will also be able to assess your knowledge about the basics of Maintenance and Reliability by taking a short test.

Introduction

- Best practices have a long history across industry.
- Best Practices consistently show superior results and should improve performance and efficiency.
- Best practices is a relative term.
- There may be barriers to adoption of best practices.

What Is a Best Practice ?

What Is a Best Practice?

A best practice is a technique, method, or process that is more effective at delivering a desired outcome than any other technique, method, or process.

Simply, a technique, method, or process may be deemed a *best practice* when it produces superior results.

Best Practice and Industrial Revolution

The notion of a best practice is not new.

Frederick Taylor (1919) said as much nearly 100 years ago:
"among the various methods and implements used in each element of each trade there is always one method which is quicker and better than any of the rest". This viewpoint came to be known as the *one best way*.

Other Best Practices History

- Henry Ford got line idea for the Model T from slaughterhouse visit.
- Taiichi Ohno, father of TPS
  TPS = Toyota Production System
  - Late 1940s/1950s visited Ford factory and a supermarket to learn mechanized system
  - (high volume/ low mix) and supply system (pull type)
  - Developed (needed) **low volume / high mix** system – today known as TPS or Kanban

Best Practice is Relative

- Best practices vary from company to company.
- For some, a practice is routine or standard.
- For others, the same practice is a Best Practice because the company is not yet producing the desired results.

Barriers to Adoption of a Best Practice*

Lack of:
- Knowledge about current best practices
- Motivation to make changes for their adoption
- Knowledge and skills required to do so

* Per American Productivity Quality Center Survey

Who is Dick Fosbury?

- In the 1968 Summer Olympics, Dick Fosbury revolutionized the high-jumping technique.
- Using an approach that became known as the *Fosbury Flop*, he won the gold medal by going over the bar back-first instead of head-first.
- Had he relied on standard practice, as did all of his fellow competitors, he probably would not have won the event. Instead, by ignoring standard practice, he raised the performance bar — literally — for everyone.

Fosbury Flop – 1968 Summer Olympic

- Fosbury's method was to sprint diagonally towards the bar, then curve and leap backwards over the bar.
- The **radical** technique quickly became the dominant technique in the event.

Didn't use Standard Practice*

* Reference Wikipedia

6 CHAPTER 1

Elements of a Best Practice

1. Practice itself – after tailored
2. Roles and responsibilities
3. Resources
4. Training
5. Information
6. Assurance - measuring performance

What Best Practices Have to Do with Maintenance and Reliability

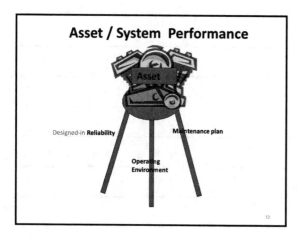

Asset / System Performance

- Designed-in **Reliability**
- **Maintenance plan**
- **Operating Environment**
- Asset

Sample Best Practices for Planning

1. **Practices**
   - All work orders are accurately detailed to the right level
     - Detailed work instructions, tools, confined space, JSA, etc.
   - Work plans include collection of condition data and work history
2. **Roles and Responsibilities**
   - Roles and responsibilities are well defined for planners
   - Others i.e. scheduler / craft supervisor know their role
3. **Resources**
   - Sufficient – adequate number of planners available
4. **Training**
   - Training program for planners available
5. **Information**
   - Planners have access to the required information for their role
6. **Assurance**
   - KPI established - measuring effectiveness of planning function

Best Practices Benchmark Data

Best of the Best Performance Measure	Best Practice Benchmark	Typical World Class
Maintenance Cost as a percent of RAV (RAV — Replacement Asset Value)	2–9%	2.0–3.5%
Maintenance - Material Cost as a percent of RAV	1–4%	0.25–1.25%
Schedule Compliance	40–90%	>90%
Percent (%) Planned work	30–90%	>90%
Production Breakdown Losses	2–12%	1–2%
Parts Stockout Rate	2–10%	1–2%

Summary

- A best practice is a technique or methodology that is found to be the most effective and has consistently shown to achieve superior results.
- When implemented appropriately, a best practice should improve performance and efficiency in a specific area.
- Best practice is a relative term. A practice may be routine or standard to some, but to others, it may be a best practice because their current practice or method is not effective in producing the desired results.
- The key barriers to adoption of a best practice are a lack of:
  - Knowledge about current best practices
  - Motivation to make changes for their adoption
  - Knowledge and skills required to make changes

1.7 Self-Assessment Questions

1. Define a best practice. What are the barriers to implementing best practices?

2. What are keys factors that impact the performance of plant machinery?

3. Why has reliability become a comparative advantage in today's environment?

4. Identify five key performance measures in the area of maintenance and reliability. Elaborate on each element of these measures. What are typical "world class benchmark number" for these performance measures?

5. Define what makes a benchmark "World Class." Discuss, using specific examples.

6. Define planned work and identify its benefits. What is a typical world class benchmark number?

Chapter 1 - Comprehension Assessment Questions

1. Who is considered one of the fathers of modern management?
2. What are some benefits of using a "best practice"?
3. What are some typical characteristics of a "best practice"?
4. Why would some say that "best practice" is a relative term?
5. Who is Dick Fosbury, and for what is he known?
6. What are benefits organizations can gain by implementing "best practices" in maintenance and reliability
7. When does a "best practice" tend to spread throughout industry?
8. What is an asset?
9. What does it mean to benchmark?
10. What is maintenance?
11. What is meant by reliability?
12. How reliability is typically expressed and calculated?
13. What does MTBF stand for?
14. What is the objective of performing better maintenance and improving reliability?
15. What is meant by inherent reliability?
16. What is meant by the operating environment?
17. What is meant by a maintenance plan?
18. What is the objective of a good maintenance plan?
19. What should a maintenance plan include?
20. Give some characteristics of M&R best practice organizations.
21. What is meant by RAV?
22. How does one calculate the measure Maintenance Cost as a Percentage of Replacement Asset Value?
23. What types of maintenance are typically included in the Maintenance Cost portion of the measure Maintenance Cost as a Percentage of Replacement Asset Value?
24. How does one calculate the measure Maintenance Material Cost as a Percentage of Replacement Asset Value?
25. What types of maintenance materials are typically included in the Maintenance Material Cost portion of the measure Maintenance Material Cost as a Percentage of Replacement Asset Value?
26. Calculate both Maintenance Cost as a Percentage of Replacement Asset Value and Maintenance Material Cost as a Percentage of Replacement Asset Value for the following data set.

CHAPTER 1

Organization	Maintenance Work Orders	Maintenance Labor Hours	Maintenance Labor Hours Cost	Maintenance Material Cost	Cost of Removing Old Assets	New Asset Purchase Costs	Cost of Intalling New Assets
A	1,384	114,872	$ 6,002,062	$ 1,500,516	$ 54,018,558	$ 300,103,100	$ 45,015,465
B	2,193,482	2,697,983	$ 140,969,604	$ 21,145,441	$ 133,622,098	$ 982,515,425	$ 157,202,468
C	350	59,150	$ 3,090,588	$ 1,236,235	$ 95,190,095	$ 288,454,833	$ 63,460,063
D	23,410	538,430	$ 28,132,968	$ 8,439,890	$ 51,866,962	$ 664,961,050	$ 86,444,937
E	928,374	5,263,881	$ 275,037,760	$ 33,004,531	$ 220,986,861	$ 2,678,628,622	$ 294,649,148

27. How does one calculate Schedule Compliance?
28. What percentage of labor hours should be included on a maintenance schedule?
29. How does one calculate Percentage Planned Work? Calculate both Schedule Compliance and Percentage Planned Work for the following data set.

Organization	Maintenance Work Orders	Maintenance Labor Hours	Maintenance Work Orders Planned	Maintenance Labor Hours Planned	Maintenance Work Orders Scheduled	Maintenance Labor Hours Scheduled
A	1,384	114,872	1,201	101,478	1,341	108,623
B	2,193,482	2,697,983	1,596,416	2,059,101	1,855,028	2,263,338
C	350	59,150	199	31,616	236	38,844
D	23,410	538,430	7,779	191,681	18,433	414,160
E	928,374	5,263,881	464,001	2,509,818	238,499	1,253,856

30. Who must be responsible for assets in order to positively affect the measure Production Breakdown Losses?
31. What is meant by the measure Parts Stock-Out Rate?
32. Regarding best practices, what is essential for its successful and effective implementation?

Chapter 1 - Research Questions
(Always cite your references when answering research questions.)

1. Why did the "Fosbury Flop" became a best practice? Provide a more detailed version of Dick Fosbury's story.
2. According to other sources, what are other improvements that an organization can make by implementing maintenance and reliability "best practices"?
3. Other than those given by the American Productivity and Quality Center (found in this book), what are other barriers to adopting "best practices"?
4. According to other sources, what factors affect asset performance?
5. According to other sources, what are other "best practice" maintenance and reliability measures?

Chapter 2
Culture and Leadership

" Effective leadership is putting first thing first.
Effective management is discipline, carrying it out."
- Stephen Covey

Chapter 2: Culture and Leadership

2.1 Introduction
2.2 Key Terms and Definitions
2.3 Leadership and Organizational Culture
2.4 Strategic Framework: Vision, Mission, and Goals
2.5 Change Management
2.6 Reliability Culture
2.7 Measures of Performance
2.8 Summary
2.9 Self-Assessment Questions
2.10 References and Suggested Reading

Chapter 2 Objectives

- Organizational culture
- Leadership and its role
- Vision, mission, and goals
- Reliability culture
- Change management and the role of change agents

Introduction

- Successful implementation of a new – best practice is a challenge
- Requires enthusiasm and a positive attitude of workforce – who impacted with the change
- Does not need fear nor distrust
- Leadership key enabler (communication)
- Strategic framework (vision, mission, etc.)
- Need to create culture of excellence – a culture of reliability

Key Factors in Making Profit

- Process – Assets
- Material
- People
  - Leadership
  - Workforce

What Is Meant By Culture?

What Is Organizational Culture?

Culture refers to an organization's values, beliefs, and behaviors. In general, it is the beliefs and values which define how people interpret experiences and behave, both individually and in groups. Culture is both a cause and a consequence of the way people behave.

Behavior and success are key enablers in creating the culture

Change and Culture

Change

⇩

Behavior

⇩

Way of Life

⇩

Change becomes a permanent part of the new culture

CULTURE AND LEADERSHIP

What Is Leadership?

Leadership*

"Leadership is leaders inducing followers to act for certain goals that represent the values and the motivations — the wants and needs, the aspirations and expectations — of both leaders and followers. And the genius of leadership lies in the manner in which leaders see and act their own and their followers' values and motivations."

*as defined by James McGregor Burns, author of the book *Leadership*

Leadership*

"Leadership is the art of accomplishing more than the science of management says is possible."

Gen. (ret) Colin Powell
Author and speaker

The Leadership Secrets of Colin Powell (Oren Harari)

Leadership*

- Get out of the office and circulate among the troops
- Build strong alliances
- Persuade rather than coerce
- Honesty and integrity are the best policies
- Never act out of vengeance
- Have courage to accept unjust criticism
- Be decisive
- Lead by being led
- Set goals and be results-oriented
- Encourage innovations
- Preach a VISION and continually reaffirm it

Lincoln on Leadership: Executive Strategies for Tough Times (Donald Phillips)

Key Attributes of Leadership

- Charisma — 7%
- Competence — 8%
- Communication — 21%
- Energizing People — 31%
- Vision — 33%

Leadership Supports:

- Creating vision/mission
- Ensuring resource availability
- Empowering people
- Individual /group goals aligned with corporate vision
- Viewing training as investment in developing people
- Aligning/integrating changes and process improvements to objectives

CULTURE AND LEADERSHIP

Leader(ship) is a key enabler in creating a culture (<u>reliability culture</u>) by providing vision – direction, resources ...

What Is A Company's Strategic Framework?

What Do You Think These Mean?

- Vision
- Mission
- Strategy
- Goals
- Action plans

CHAPTER 2

Strategic Framework

```
        /\
       /VISION\
      /MISSION \
     /STRATEGIES\
    /   GOAL    \
   / ACTION PLANS \
  /_____\
```

Strategic Framework

- **Vision** — Where are we and the organization going?
- **Mission** — What we are planning to do? What will be accomplished?
- **Values** that shape our actions — Why are we going in this direction?
- **Strategies** that zero in on key success approaches — How will we get there?
- **Goals** and **Action Plans** to guide our daily, weekly, and monthly actions — When will we get there?

Vision

A vision statement is a short, succinct, and inspiring declaration of what the organization intends to become or to achieve at some point in the future.

CULTURE AND LEADERSHIP

Where Do We Want Our Maintenance Organization to Be?

World's Best Maintenance & Reliability Program

Mission

A mission statement is an organization's vision translated into written form. It's the leader's view of the direction and purpose of the organization. For many corporate leaders, it is a vital element in any attempt to motivate employees and to give them a sense of priorities.

A mission statement should be a short and concise statement of goals and priorities.

Strategy

Strategy is a very broad term which commonly describes any thinking that looks at the bigger picture. A successful strategy adds value for the targeted customers over the long run by consistently meeting their needs better than the competition does.

Strategy is a plan based on the mission an organization formulates to gain a sustainable advantage over the competition.

Objectives

- Focused on a result, not an activity
- Consistent
- Specific
- Measurable
- Related to time
- Attainable

Setting Goals

- Major outcome of strategic planning and setting objectives
- After getting all necessary information
- Should be based on its vision and mission
- Goal = Specific and realistic long-range aim for specific period
- Long-range goals set through strategic planning translated into activities to ensure reaching goal through operational planning

How Would You Answer for M&R?

- **Vision** — Where are we and the organization going?
- **Mission** — What we are planning to do? What will be accomplished?
- **Values** that shape our actions — Why are we going in this direction?
- **Strategies** that zero in on key success approaches — How will we get there?
- **Goals** and **Action Plans** to guide our daily, weekly, and monthly actions — When will we get there?

CULTURE AND LEADERSHIP

What Is Change Management?

How Does One Manage Change Effectively?

Change

"It is not always the biggest or the strongest that will survive. The one that will survive is the one that is able to *change* and *adapt* to the environment."

Change

- Change is constant
- Change is a characteristic of a healthy organization
- Change keeps organizations competitive in global market
- If you fail to change, your competition will change without you
- And you will…

Culture Change Process

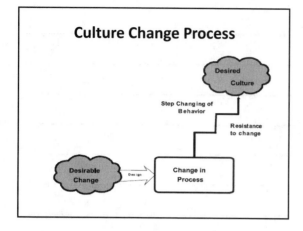

Changing Culture

1. Influencing the behavior to change

2. Overcoming resistance to change

Influencing Behavior

- Increase understanding (i.e., why the change is needed and how it relates vision)
- Goal setting and expectations
- Praise, rewards, and celebration
- Roles definition and clarification
- Procedures and standards
- Persistence, tenacity, and discipline

Overcoming Resistance

- Listen and communicate
- Create awareness
- Educate and train to create understanding
- Get team members involved and let them see some success
- Empower team members to improve, tailor the process — change if needed

Change Agents

- Very important role in implementing reliability (or other) culture change
- Have clout, conviction, charisma, resourcefulness, and other innate and trained skills
- Make things happen
- Keep others engaged in implementation
- Role should be filled by senior management person who has respect and trust of people

What Is a Reliability Culture?

Maintenance & Reliability Excellence Culture

- Is it...
  - ✓ Lowest Downtime ?
  - ✓ Highest Uptime ?
  - ✓ Zero breakdowns ?
  - ✓ Zero accidents ?
  - ✓ Highest OEE ?
  - ✓ Best scheduling compliance ?
  - ✓ Lowest maintenance cost to ERV/RAV ratio, etc...

  Yes...All of the above

Bottom Line

Supporting plant / organization's business proposition by <u>minimizing maintenance-related costs</u> while <u>maximizing the capacity to produce</u> products /services

... Excellence is a moving target.

Reliability Culture Scenarios

See pages 39-42

Results of Reliability Culture

- Prevention of failures becomes an emphasis at every organizational level.
- The entire workforce (operators, maintainers, engineers) is focused on asset reliability.
- Assets are available to produce (or provide service) when needed
- Assets are maintained at a reasonable (optimal) cost
- An effective and optimized facility maintenance plan in place -- RCM/CBM based

Results of Reliability Culture (cont'd)

- 80/20 principal applied to prioritize the work.
- Most of the work is planned and scheduled.
- If an asset fails, it gets fixed quickly, the root cause is determined, and action is taken to prevent future failures.
- Facility/asset reliability analysis is performed on a regular basis to increase uptime.
- The workforce is trained and taught to practice reliability-based concepts and best practices on a continuous basis.

What Performance Measures Could Be Used in the Area of Culture and Leadership?

Measures of Performance

- Performance measures in Leadership can be classified into four categories:
  - People Initiatives
  - Collaboration Initiatives
  - Organizational Initiatives
  - Professional Success Indicators

Summary

- Business (Corporate) success depends on:
  - the vision articulated by the organizational leaders and management
- Culture refers to an organization's values, beliefs, and behaviors
  - The most effective visions are those that inspire, usually asking employees for the best, the most, or the greatest.
  - A vision explains where the organization wants to be and is an important element of creating a reliability culture
- Organizations with strong reliability cultures achieve higher results because employees sustain focus both on **what** to do and **how** to do it.

Summary (cont'd)

To sustain a reliability culture, the reliability / maintenance leaders in an organization must continue to provide the right tools, training, and education to both the operators and maintainers together as a team. They need to ensure that the workforce is always current on:

- Knowledge — of best practices
- Teamwork — to assure communication and understanding
- Focus — on the right goals for business success
- Planning — to create a roadmap for knowing where they are and where they want to be
- Processes — documentation, adherence, and discipline
- Measurements — to provide feedback and control and to ensure that they, the leadership, continue to support the continuous improvement environment and are creating a conducive, sustainable culture

2.9 Self - Assessment Questions

1. What are the key attributes of a leader?

2. Why is vision important? Also define vision statement.

3. Define MBWA. Why is it considered one of the key leadership practices?

4. Define an organizational culture.

5. What are the key benefits of having a mission statement?

6. Why are mission and vision statements important for an organization?

7. Define reliability culture.

8. Why is change management an important part of creating the right reliability culture?

9. Define the role of a change agent. Who is best qualified to perform this role?

10. State the difference between a manager and leader

Ch 2 Comprehension Assessment Questions

1. Why do reliability organizations with strong work cultures achieve higher results?
2. What are two key enablers in creating culture?
3. Describe the circular flow of mutual causation among organizational behavior success in culture.
4. Describe leadership according to James Macgregor Burns.
5. Describe leadership according to retired Gen. Colin Powell.
6. How are the leaders and organization similar to the definition of leader as it applies to a plant?
7. Give at least three qualities that successful leaders have that are needed to improve their processes.
8. Describe the strategic framework for success.
9. Give an example maintenance vision statement.
10. What questions should the mission statement answer?
11. Why are mission statements not often taken seriously in organizations?
12. What is a strategy?
13. How does a successful strategy add value for the targeted customers over the long run?
14. What must objectives be in order to be successful?
15. What is the major outcome of strategic planning?
16. In order to succeed in a culture change, what must an organization have?
17. What are the key elements of any successful effort to change culture?
18. What actions are suggested to influence the behavior to change?
19. What actions are suggested to overcome the resistance to change?
20. What skills do change agents need to have?
21. Describe the characteristics of an organization with a strong reliability culture.
22. What does it take to change an existing culture of run-to-failure and little or no PM program to a sustainable reliability culture?
23. What are the differences between the different reliability cultures discussed at the end of this chapter
24. What qualities are leaders evaluating on with the LPI measurement model?
25. What behaviors are leaders assessed by according to the LPI measurement model?
26. What categories can performance measures in leadership be classified into?
27. What factors do People Initiatives include?
28. What factors do Collaboration Initiatives include?

29. What factors do Organizational Initiatives include?
30. What factors do Professional Success Indicators include?
31. What are the typical critical skills and competencies for executives?
32. What is the typical knowledge that a senior leader should have?
33. What is the typical experience that leadership might require?
34. What are the key attributes that successful leaders should demonstrate?
35. In order to sustain a reliability culture, what must the reliability and maintenance leaders and organization continue to ensure their workforce remain current on?

Research Questions

(Always cite your references when answering research questions.)

1. Research more recent survey results about leader characteristics. Discuss their meaning as well as their similarities and differences from the survey results presented in this chapter.
2. Describe one or more of the more leadership principles and how they compare to one of the leadership principle groups in this chapter.
3. Based on the Strategic Framework for Success triangle (Figure 2.4), where would the values element of strategy (as discussed on the same page) play a part? Support your answer with research.
4. Choose at least five vision statements and/or mission statements not included in this chapter and evaluate why they are or are not "good" statements.
5. What are the more recent views about vision statements and mission statements?
6. Describe at least one strategic planning process and explain how this could be modified to apply to maintenance and reliability.
7. Describe the strategic planning process and how setting goals fits into this process.
8. Compare the advantages and disadvantages of a formal (written-down) change management process (supported by research).
9. Discuss at least one change management process model.
10. Evaluate more recent views of the role of a change agent in an organization.
11. Describe more recent views of the innate, learned, and experienced characteristics of a change agent.
12. Compare at least two organizations in their reliability cultures.
13. Describe any formal performance measures being used by organizations in the areas of culture and leadership.

Chapter 3
Understanding Maintenance

"Your system is perfectly designed to give you the results that you get"

- - W. Edwards Deming

Chapter 3 Understanding Maintenance

- 3.1 Introduction
- 3.2 Key Terms and Definitions
- 3.3 Maintenance Approaches
- 3.4 Maintenance Practices: Others
- 3.5 Maintenance Management System: CMMS
- 3.6 Maintenance Quality
- 3.7 Maintenance Assessment and Improvement
- 3.8 Summary
- 3.9 Chapter Assessment
- 3.10 References and Suggested Reading

Chapter 3 Objectives

- Why do maintenance?
- The objective of maintenance
- Benefits of maintenance
- Types of maintenance approaches
- Purpose of CMMS/EAM
- Maintenance quality challenges
- Importance of assessing and improving your maintenance program regularly

Introduction

- Maintenance concerned with keeping an asset in good working condition so the asset may be used to its full productive capacity.
- Maintenance function includes:
  - Upkeep
  - Repairs

Definitions of Maintenance

- Simple Definition:
  - the work of keeping something in proper condition; upkeep.
- Broader Definitions:
  - Keep in "designed" or acceptable condition
  - Keep from losing partial or full functional capabilities
  - Preserve or protect

Maintenance Problems of the Past

Literature related to maintenance practices over past few decades indicates that most companies did not commit the necessary resources to maintain assets in proper working order.
- Assets were allowed to fail
- Then the necessary resources were committed to repair or replace failed assets or components

Maintenance Viewpoints of the Past

- Maintenance function was viewed as the necessary evil

- Therefore maintenance did not receive the attention it deserved

Things Are Changing

- In last few years, this practice has changed dramatically.
- Corporate world has begun recognizing the reality that maintenance does add value.
- Very encouraging to see maintenance moving from so-called "backroom" operations to the corporate board room

Things Are Changing – Case in Point

- 2006 annual report for Eastman Chemical included a couple of slides related to maintenance and reliability
- Stressing company's strategy of increasing equipment availability by committing adequate resources for maintenance.

New Paradigm of Maintenance

- Shift thinking to capacity assurance
- With proper maintenance, capacity of an asset can be realized at designed level
- Acceptable capacity level = target capacity level set by management (no more than the designed capacity)
- Management could justify increased maintenance cost by reducing downtime to increase capacity to design level

What Is Meant By a Maintenance Approach?

Maintenance Approaches

- Companies have many different approaches when it comes to their maintenance programs
- All approaches have at their basis the requirement to keep their facility's assets at whatever capacity level is necessary for their current operational needs

Maintenance Approaches (cont'd)

- Some of these maintenance programs are more structured than others (e.g., RCM analysis or annual/multi-year plan).
- The truth is that a company will have a maintenance program whether they admit it or not.
- Their program will simply be more costly than it has to be because they will live in a reactive maintenance state.

Why Have a Structured Maintenance Program?

1. Should improve production capacity
2. Should reduce overall facility costs

Structured Maintenance Program

- Reduces production downtime — the result of fewer asset failures
- Increases life expectancy of assets, thereby eliminating premature replacement of machinery and asset
- Provides more economical use of maintenance personnel
  - Due to working on a scheduled basis
  - Instead of unscheduled repair of failures
  - Also reduces overtime costs

UNDERSTANDING MAINTENANCE

Structured Maintenance Program (cont'd)

- Reduces cost of repairs by reducing secondary failures. When parts fail in service, they usually damage other parts
- Reduces product rejects, rework, and scrap due to better overall asset condition
- Identifies assets with excessive maintenance costs, indicating the need for corrective maintenance, operator training, or replacement of obsolete assets
- Improves safety and quality conditions

What Are Some Different Maintenance Approaches?

Basic Maintenance Philosophy (Approach)

Really only two-fold (one of two choices) for each asset:

Do some form of maintenance
to prevent failure of an asset

OR

Allow asset to run-to-failure

CHAPTER 3

Basic Maintenance Approaches

- Condition Based Maintenance (CBM) / Predictive Maintenance (PdM)
- Preventive Maintenance (PM)
- Proactive Maintenance
- Corrective Maintenance (CM) / Run-to-Failure (RTF)

  - *More details about each in subsequent chapters*

CBM/PdM Definitions

- CBM and PdM often used synonymously
- As seen by the similarity of their definitions:
  - CBM: Maintenance based on actual condition (health) of an asset as determined from non-invasive measurements and tests.
  - PdM: Maintenance based on actual condition (health) of an asset as determined from non-invasive measurements and tests. Condition measured using condition monitoring, statistical process control, or performance, or use of the human senses.

CBM/PdM Process

- Evaluates condition of assets by performing condition monitoring
  - Periodic route collection of asset health data
  - Continuous monitoring of asset health data
- Inspections mostly performed while the asset is operating, thereby minimizing disruption of normal system operations

CBM/PdM Process (cont'd)

"Predictive" component stems from goal of predicting the future trend of asset condition
- Using principles of statistical process control, trend analysis, and preselected thresholds
- To determine at what point in the future maintenance activities should be scheduled
- Allowing corrective actions to be optimized through proactive planning and scheduling
- Allowing preventive actions to be optimized by avoiding traditional calendar or run-time directed schedules

Multiple Condition Data Methods

- Vibration analysis
- Oil analysis
- Infrared (IR) thermography
- Ultrasonic — airborne, contact, etc.
- Electrical — motor, resistance, etc.
- Partial discharge and Corona detection Shock Pulse Method (SPM)
- Operational performance data — pressure, temperature, flow rates, etc.

The Challenges of CBM/PdM

- Starting a full-blown program utilizing all these technologies can be quite expensive
  - Test equipment for these different technologies
  - Training personnel to use this equipment and the technologies themselves
  - Choose for most "bang for your buck"
  - Reason to have RCM-basis for choosing where to apply each technology
- How the CBM/PdM team is organized
  - Centralized, dedicated team good way to start
  - Helps in standardizing methods and practices

Overcoming CBM/PdM Challenges

Requires firm management and organizational commitment to make program work
- Continue to push forward with financial and organizational investment
- Support proactive repairs and scheduled downtime in a timely manner

Ultimate Goal of CBM/PdM

- To identify proactive maintenance actions when:
  - Maintenance activity is most cost-effective
  - Before asset fails in service
- Adoption can result in substantial cost savings and higher system reliability:
  - Reduction in maintenance costs: 15–30%
  - Reduction in downtime: 20–40%
  - Increase in production: 15–25%

PM Definition

Maintenance at a fixed interval for inspection, component replacement, and/or overhauling
- Regardless of current condition
- Usually scheduled inspections performed to assess the condition of an asset
- Replacing service items, e.g., filters; oils, and belts and lubricating parts are a few examples of PM tasks
- May require another work order to repair other discrepancies found during PM

PM Process

- Maintenance or operations personnel regularly visit assets to assess its condition
  - Based on calendar time
  - Based on asset runtime
  - Based on asset cycles
- Checklists and procedures with task details can be helpful
  - Indicating what to check or what data to take
  - Document abnormalities other findings
  - Should be corrected before developing into failures

Goals of Performing PM

- Next best thing to CBM/PdM
- Maybe only approach with certain types of failures and/or specific equipment
- Regulatory requirements may force some level of PM
- Basic objective: To take a look at asset to determine if any telltale signs of failure or imminent failure
- If abnormalities are not corrected before they turn into failures, the PM program does not add any value

Proactive Maintenance

Three different definitions exist in industry:

Any maintenance work that has been identified in advance, planned, and scheduled

OR

All maintenance work completed to avoid failures or to identify defects that could lead to finding failures, including PdM and PM

OR

Only those tasks that are generated based on what is found during CBM and PM tasks

Corrective Maintenance (CM)

- CM is an action initiated as a result of an asset's observed or measured condition before or after functional failure.
- CM is also called Repair Maintenance
  - Because it corrects deficiencies
  - Because it allows the asset to work again after it has failed or stopped working
- Corrective Maintenance can be:
  - Scheduled Repairs
  - Major Repairs/Projects
  - Reactive Repairs (Breakdown)

What Is Meant by a Failure?

- Definition of Failure: Loss of an asset's ability to perform its required function; does not require asset to be inoperable; could also mean reduced speed or missing operational or quality requirements.
- When asset breaks down, it fails to perform its intended function and disrupts scheduled operation.
- This functional loss (partial or total) may result in defective parts, speed reduction, reduced output, and unsafe conditions.
- Function-disruption or reduction failures not given due attention will soon develop into asset stoppage.

Run-to-Failure

- Given the discussion thus far, why would I ever choose run-to-failure?
  - When analysis has been performed indicating the most cost-effective strategy to be run-to-failure
  - Because the total cost of maintenance is less than the corrective maintenance necessary for this run-to-failure strategy
  - Assuming that there is no safety impact to this run-to-failure strategy
- This should be a proactive decision of the organization for only those assets selected
  - NOT an overall organizational approach!

What Are Maintenance Practices?

Maintenance Practices

- Maintenance practice is a specific approach, not of the overall maintenance program, but of the way this program is executed
- Example maintenance practices:
  - Operator/Maintainer relationships/involvement
  - Procedures and task instructions
  - Cleaning and lubricating
  - Planning and scheduling
  - Skills development and training
  - Analysis tools and techniques
  - Designing for Reliability and Maintenance

Operator Involvement Practice

- Operator-Based Maintenance (OBM or TPM)
- Foundation piece of TPM philosophy
- Operators are the first line of defense
- Main objective is to equip operators with:
  - Ability to detect abnormalities
  - Ability to correct minor abnormalities and restore function, if they have the ability
  - Ability to set optimal asset conditions
  - Ability to maintain optimal equipment conditions

CHAPTER 3

What Is a CMMS?

CMMS

CMMS: Computerized Maintenance Management System (software)
- Software system that keeps record and tracks all maintenances activities; usually integrated with support systems such as inventory control, purchasing, accounting, manufacturing, and controls maintenance and warehouse activities.
- Other similar electronic systems
  - EAM : Enterprise Asset Management
  - ERP: Enterprise Resource Management
- Many different software vendors available

CMMS Capabilities

- Many different capabilities exist, dependent on software vendor
- Only few of the capabilities listed below:

Assets	Work Orders	PMs
Job Plans	Crew Schedules	Reports
Reliability Analysis	Configuration Management	Inventory / Spares
Work routing	Work Backlog	Material

Many more capabilities listed in book

Barriers to CMMS Acquisition

- Organization is too small for a CMMS
- CMMS project payback or savings inadequate
- MIS or IT doesn't give CMMS high enough priority
- MIS and maintenance speak different technology languages
- Participants fail to reach consensus

Why So Many CMMS Projects Fail

- Selecting wrong CMMS system for your business
- Employee turnover
- Lack of adequate training during implementation
- Employee resistance
- Being locked into restrictive hardware/software
- Inadequate supplier support for the CMMS
- High expectations and quick return on investment
- Internal Politics — Financial or IT — heads the CMMS/EAM implementation team

NOTE: CMMS projects should be led by senior maintenance /reliability management person

Selecting the "Right" CMMS

Selecting the "right" CMMS is crucial to successful implementation
- Depends upon multiple things:
  - System Features
  - Ease of Use
  - System Affordability
  - Vendor Support

What Is Meant By Maintenance Quality?

Maintenance Quality

- It is said that "Accidents do not happen, they are caused."
  - The same is true for asset failure.
  - Assets fail due to basically two reasons: poor design and human error.
- All maintenance work involves some risk.
  - Risk refers to potential for inducing various defects while performing maintenance tasks.
  - Human errors made during the PM, CBM, and CM tasks eventually may lead to additional failures from maintenance performed.

Poor Maintenance Quality Examples

During a PM task you may cause damage during inspection, repair adjustment, or installation of replacement part by:
- Installing material/ part that is defective
- Incorrectly installing a replacement part, or incorrectly reassembling
- Reintroducing infant mortality by installing new parts which have not been tested
- Causing damage to an adjacent asset or component during a maintenance task

Improving Maintenance Quality

To create high quality and motivated personnel, try the following measures:
- Provide training in best practices and procedures for specific assets.
- Provide appropriate tools to perform tasks effectively.
- Get involved in performing FMEA/RCA/RCFA, and in developing maintenance procedures.
- Follow up to assure quality performance and to show that management cares for quality work.
- Publicize reduced costs with improved uptime, the result of effective maintenance practices.

Why Would We Assess Our Maintenance Program?

Maintenance KPIs

- KPI: Key Performance Indicator
- KPIs (also called metrics) are an important management tool to measure performance (to assess our maintenance program)
    - "What gets measured gets done."
- KPIs also help us make improvement actions
    - "If we can't measure it, we can't improve it."
- However, too much emphasis on KPIs, including the wrong KPIs, not right approach
    - Selected indicators shouldn't be easy to manipulate just to "feel good."

Maintenance KPI Selection

The following criteria are recommended for selecting the best KPIs:
- Should encourage the right behavior
- Should be difficult to manipulate
- Should be easy to measure — data collection and reporting

Example Maintenance KPIs and Their Benchmark Values

Metric	Typical	World Class
Maintenance Cost % of ERV	3 - 9 %	2.5 - 3.5 %
Production Loss - Breakdowns	5 - 10 %	< 1 %
Reactive - CM Unscheduled	40 - 55 %	< 10 %
Planned Maintenance	40 - 70 %	85 - 90 %
Overtime	10 - 20 %	< 5 %
Rework - Maintenance Quality	~ 10 %	< 1 %

How Can We Improve Our Maintenance Program?

Maintenance Improvement

Maintenance effectiveness can be improved
- By optimizing the maintenance work tasks
    - These tasks can be optimized by using tools and techniques such as FMEA, RCM, and predictive technologies
- By effective task execution through the utilization of the many tools available
    - These tools and techniques can help optimize the content of the work tasks to be accomplished.

Finding the "Sweet Spot" of Maintenance

Finding the Sweet Spot

- Maintenance Cost Change =
    [Proposed Maint Cost] – [Current Maint Cost]
- Profit Change = [Profit Per Unit] x [Production Change]
- Production Change =
    [Proposed Production] – [Current Production]
- Production =
    [Production Per Hour] x [Actual Production Hours]
- Actual Production Hours =
    [Available Production Hours] x [% Uptime]
- Available Production Hours =
    [Max Production Hours] – [Scheduled Downtime Hours]

Create a "Living Maintenance Program"

- Continually review processes, procedures, and tasks for applicability, effectiveness, and interval frequency
- Standardize procedures and maintain consistency
- Identify and execute mandated tasks to ensure regulatory compliance
- Apply and integrate new predictive technologies
- Ensure task instructions cover lockout / tag out procedures and all safety requirements
- Ensure operation and maintenance personnel understand importance of PM practice and provide feedback for improving PMs

Summary

- Maintenance has recently shifted from merely maintaining and repairing assets to a new paradigm of capacity assurance.
- Benefits of a structured maintenance program are improving production capacity and reducing overall facility costs.
- Maintenance can use multiple approaches:
  - Condition Based/Predictive Maintenance
  - Preventive Maintenance
  - Proactive Maintenance
  - Corrective Maintenance / Run-to-Failure

Summary (cont'd)

- Multiple best maintenance practices can be used, including operator-based maintenance.
- A CMMS is an important tool to support maintenance activities.
- Another important piece of a maintenance program should be to assess and improve your maintenance program continuously.

3.9 Chapter Self-Assessment Questions

1. Define maintenance and its role.

2. What are the different categories of maintenance work?

3. What can equipment operators do to support maintenance?

4. Why would an organization support operators getting involved in maintenance?

5. Why would an organization need to have a CMMS? What is the difference between a CMMS and an EAM?

6. List five maintenance metrics and discuss why they are important.

7. Name five PdM technologies and discuss how they can help reduce maintenance costs.

8. Define Proactive Maintenance

9. What are the benefits of a structured maintenance program?

10. How can a CMMS/EAM system help improves maintenance productivity?

Chapter 3 Comprehension Assessment Questions

1. What new name can be given to maintenance function? Discuss the benefits of name selected

2. Based on the following information, calculate the improved capacity from maintenance improvements.

   Consider an asset with production capacity of 100 units per hour with its product demand able to handle all the units that could be produced. Each unit adds additional profit of $5 per unit. The current maintenance program for this asset requires scheduled downtime of 500 hours per year at a cost of $25K, yielding 80% asset uptime. With a change in maintenance approach, the asset will require 600 hours of downtime per year at a cost of $50K (including labor & additional equipment cost). This will increase asset uptime to 85%.

   Should the company make this improvement to their maintenance program?

3. What does literature indicate about maintenance practices over the past few decades?
4. How was the maintenance function viewed in the past
5. What is the basis of all maintenance approaches?
6. What is the consequence of failing to perform maintenance activities?
7. What is the most important reason to have a structured maintenance program?
8. What is the goal of condition based maintenance (CBM)/predictive maintenance (PdM)?
9. At what point of during asset operation are CBM/PdM inspections typically performed?
10. What is a benefit of performing CBM/PdM inspections at the point of asset operation that they are performed? Minimizing disruption of normal system operation (p. 55)
11. What are the benefits of adoption of a CBM/PdM program in the maintenance of an asset?
12. What is the basic difference between a CBM/PdM and Preventive Maintenance (PM) program?
13. Give an example that shows the difference between a CBM/PdM and PM approach to maintenance.
14. What have past studies shown that a well implemented CBM/PdM program can provide as far as savings when compared to a traditional PM program?
15. What are the typical savings that you will see with a fully implemented CBM/PdM program when there is no good PM program in place?
16. What are some of the challenges of starting a full-blown CBM/PdM program utilizing all possible technologies?
17. What are two reasons why a PM approach would be the best maintenance approach?
18. Describe the basic approach when it comes to doing PM.
19. What are two different time-based approaches for a PM program?
20. When is it appropriate to select a run-to-failure strategy for an asset rather than a PM or CBM/PdM strategy?
21. What are the main objectives of a PM approach of maintenance?
22. Calculate the proactive maintenance ratio metric based on the following data.

Total Preventive Maintenance Cost	$250K
Total Condition Based Maintenance/ Predictive Maintenance Cost	$150K
Total Maintenance Cost Identified by PM & CBM/PdM	$400K
Total Corrective Maintenance Cost	$200K

23. What is the basic definition of corrective maintenance?
24. What are the three major classifications of corrective maintenance?
25. What are the potential outcomes of a functional failure?
26. What are some potential asset abnormalities that are indications of imminent trouble for asset?
27. At what point during operation has it been observed that a high percentage of failures are likely to occur?
28. What is the responsibility of operators with regard to maintenance?
29. What is the main objective of an Operator-Based Maintenance program?
30. What are the asset-related skills that are necessary for an Operator-Based Maintenance program to be successful?
31. What is TPM?
32. What are some maintenance skills that operators should be able to learn as part of maintenance skills training?
33. What are operators able to do using sensory tools to identify problem areas with assets?
34. What are suggested measures to help in achieving the goal of Total Productive Maintenance?
35. What are some maintenance issues that cleaning will allow easy identification of?
36. What are some potential lubrication issues?
37. List at least 10 capabilities that a CMMS system should perform.
38. What are the major features the work order management module of a CMMS should include?
39. What are some examples of customization capability for a CMMS?
40. What are the benefits of tracking life cycle costs of assets?
41. What is the one of the more important trends being adopted into the CMS industry?
42. What are the typical barriers to CMMS acquisition?
43. What are the three major facets to be considered in selecting the right CMMS for your organization?
44. What are some of the reasons why a CMMS project may fail?
45. Who should ideally lead the CMMS implementation team?
46. What are the major steps that should be followed in deciding to acquire a CMMS?
47. What are the two basic reasons that assets fail?
48. Why are most asset failures deemed to be preventable?
49. What are some examples of damage that may occur due to human error during maintenance?

58   CHAPTER 3

50. What measures are suggested to create high quality and motivated personnel?
51. What does the acronym KPI stand for?
52. What are two benefits of maintenance KPIs?
53. What criteria are recommended for selecting the best KPIs?
54. What are some key maintenance metrics and their associated benchmark data (typical findings and world-class values)?
55. How can maintenance effectiveness be improved?
56. What are the key elements that make up a living maintenance program?

Research Questions
(Always cite your references when answering research questions.)

1. Create a different categorical breakdown of maintenance activities than what is described in this chapter and discuss why this categorical breakdown is better, worse, or equivalent value in describing an organization's maintenance activities.
2. Describe a less commonly used CBM/PdM technology (including the estimate cost to implement this technology, how to use (or the process for using) this technology, the types of assets and failure modes for which the technology can be used, and the limitations of this technology).
3. Evaluate at least one expert's view of each of the three different definitions applied to the term "Proactive Maintenance" and then explain which definition you prefer and why.
4. Discuss why or when run-to-failure would be the most appropriate maintenance approach/strategy.
5. Describe a maintenance practice not covered in detail in this chapter and why this activity may or may not be more important than the one covered in detail in this chapter.
6. Explain why/when an organization should not use a CMMS.
7. Contrast two different CMMS/EAM packages; select which you think is the better of the two and explain why.
8. Compare the advantages, disadvantages, and potential challenges of implementing mobile technology for an organization's CMMS.
9. Explain why human error is or is not prevalent in maintenance activities using research outside this book.
10. Discuss ways that organizations are using to limit their risk with regard to human error in maintenance.
11. List key maintenance KPIs (at least 5) not discussed in the book and provide your information sources.
12. Evaluate at least three improvement processes/practices for improving a maintenance program.
13. Explain why many CMMS/EAM systems don't provide projected benefits?
14. Describe the quality issues in maintenance. Discuss what we could do to improve.

Chapter 4
Work Management: Planning & Scheduling

A goal without a plan is just a wish.
— Antoine de Saint-Exupery

Chapter 4: Work Management

4.1 Introduction
4.2 Key Terms and Definitions
4.3 Work Flow and Roles
4.4 Work Classification and Prioritization
4.5 Planning Process
4.6 Scheduling Process
4.7 Turnarounds and Shutdowns
4.8 Measures of Performance
4.9 Summary
4.10 Self Assessment Questions
4.11 References and Suggested Reading

Chapter 4 Objectives

- The basic work flow process
- The different roles used (planner, scheduler, etc.) in managing work
- Work classification and prioritization
- The importance of backlog management
- Why planning is necessary
- The planning process
- Why scheduling is necessary
- The scheduling process
- Turnaround management

WORK MANAGEMENT: PLANNING AND SCHEDULING

Introduction

Previously covered the development of the proper maintenance tasks to keep our assets working
- This translates to an effective maintenance program
- An important element to a better overall maintenance program
- One that ensures the new paradigm of maintenance is sustained (plant capacity)

Introduction (cont'd)

Maintenance tasks should also be performed efficiently to be sustained cost effectively.
- To truly reduce overall operations and maintenance costs, these tasks must be executed efficiently and effectively
- This means that maintenance personnel must be more productive
- This is achieved by eliminating or minimizing avoidable delays and wait time.

What Does It Mean for Maintenance Craft to Be Productive?

The Unfortunate Truth

Industry reports low productivity levels in maintenance craft groups worldwide
- Major M&R conferences have indicated maintenance craft productivity 30–50%
- 3–4 hours average productive within 8-hr shift
- Some call this productive time "wrench time"
- Time craft personnel actually spend their efforts repairing the assets
- As opposed to receiving unclear instructions, going to store for the right tools, waiting for other craft to arrive or release asset, etc.

One View of Maintenance Productivity

- There are some managers who say:
  - They would be thrilled to hear that their maintenance craft workers are sitting idle
  - Spending most of their time waiting for breakdowns to happen
- The "Maytag Repairman" image:
  - Maintenance groups often compared to a fire department
  - Where fewer fires to battle are better

How Should Be Viewed

A maintenance group can be far more productive in so many ways
- Becoming proactive instead of responding to emergencies like fire departments.
- Performing PM & CBM tasks
- Participating in process improvements
- Working on capital improvement initiatives
- Upgrading their skills, training others, and educating operators to run the assets properly to minimize errors

WORK MANAGEMENT: PLANNING AND SCHEDULING

To Be Productive Means To Be Prepared

The Way of the Unprepared

- Imagine yourself repairing a leaky faucet or dishwasher at home.
- You have been asked repeatedly to fix it, and finally you find the time.
- Can you recall the number of times you:
  - Went back and forth to the garage, to the tool box, or to the hardware store?
  - To get the correct-sized tool or material?
- Probably took 4 hours or more to complete this task.

What Does It Mean to Be Unprepared?

- Unprepared = Work Not Properly Identified, Planned, and/or Scheduled
- Not knowing exactly what needs to be done = Unprepared (Not Properly Identified)
- Disorganized work activity = Unprepared (Not Properly Planned)
- Not having the right parts or tools = Unprepared (Not Properly Planned)
- Not having proper work task instructions = Unprepared (Not Properly Planned)
- Frequent work interruptions and restarts = Unprepared (Not Properly Scheduled)

The Way of Lincoln

- Abraham Lincoln could be considered to practice the way of the prepared
- He understood the value of preparation before starting a task.
- He is credited with the following quote:
    *"If I had eight hours to cut a tree,
    I'd spend six hours in sharpening the axe."*

The Way of the Prepared

- Next time a plumber is called
    - The plumber comes in and assesses the problem
    - Goes back to the truck & gets the right tools and parts
    - Corrects the problem in 40–45 minutes
- Maybe you could do it in 2 hours if you had the right tools and right parts?
- The point is proper preparation with the right tools, parts, and instructions can save time and avoid wasteful activities.

Where Work Management Helps

- Effective work management helps to be more prepared for the maintenance task
- Work management ensures plant capacity is sustained cost-effectively by avoiding delays & minimizing non-productive work
    - Minimizes wait time & other wasteful activities (including id of the "right" work)
    - Effectively planned and scheduled jobs take substantially less time than unplanned jobs
    - One estimate is that every hour invested in planning saves 1-3 hours in work execution

WORK MANAGEMENT: PLANNING AND SCHEDULING

Effective Work Management

- Defining and clarifying the right work
- Prioritizing work
- Developing the work sequence and steps to complete the task
- Identifying necessary tools, materials, and skills sets (work roles)
- Ensuring on-schedule availability of materials and assets
- Scheduling the work to be done with agreement from production on scheduled time
- Ensuring details of completed work are documented in CMMS

What Is Meant by Work Flow?

What Are the Different Roles That Add Value in the Work Flow?

Maintenance Approaches

- Recall Basic Discussion in Chapter 3 on Different Maintenance Approaches
    - CBM/PdM
    - PM
    - Proactive Maintenance
    - CM/RTF
- For Work Flow, This Translates to These Categories:
    - Preplanned Work (PM/CBM/PdM)
    - New Work from PM/CBM/PdM
    - Reactive Work (Breakdown/Emergency)

Chapter 4

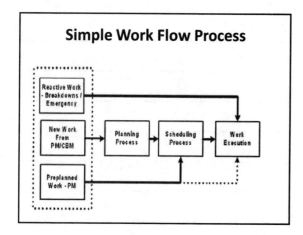

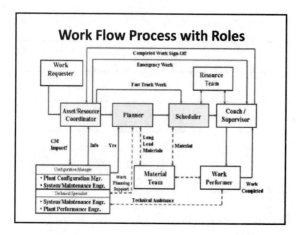

How Should Work Be Classified?

How Should Work Be Prioritized?

Work Classification

- As shown in previous work flow diagrams, work classification plays a part in work flow
- It really does not matter how work is classified as long as the CMMS provides data in a format to make good decisions and the classification aids in work flow
- One method of work classification would be the same as previous slide on maintenance approaches
- But there are other work classifications, dependent on organization needs & metrics

Other Work Classifications

- Preventive Maintenance (PM):
  - Calendar-based (Age-Related)
  - Run-based (Usage-Related)
  - Condition-based (Health-Related)
  - Operator-based maintenance
- Corrective Maintenance (CM):
  - CM Routine work resulted from PMs
  - CM Major Repairs/Projects
  - CM Reactive (Breakdown/Emergency)

Work Prioritization

- Work prioritization is another classification of work useful to the work flow process
- Work priority codes allow ranking of work to accomplish work in order of importance
  - Will eliminate tasks being done "on a whim"
  - Instead allow work to be accomplished according to its true impact on the overall operations
  - Allow maintenance delivery function to be executed in a far more effective manner with better plant-wide utilization of resources

Work Prioritization (cont'd)

- Too many organizations neglect the benefits of a clearly-defined prioritization system.
- Many organizations have more than one prioritization systems – most ineffective.
- Drawbacks of Not Clearly Defining Priorities
  - Wasted maintenance man-hours on tasks of low relative importance
  - Critical tasks being lost in maintenance backlog
  - Dissatisfied operations customers
  - Lack of faith in the effectiveness of the maintenance delivery functions

Work Prioritization Calculations

- Work priority is a combination of asset criticality and work impact
- Work Priority = Asset Criticality x Work Impact
  - WO #1: Asset Criticality of 5 & Work Impact of 4 gives an overall job priority of 20
  - WO #2: Asset Criticality of 4 & Work Impact of 4 gives an overall job priority of 16
  - In this case, WO #1 will have the higher priority when compared to WO #2

Example Work Priority Elements

- Asset Criticality

5	Critical safety-related items and protective devices
4	Critical to continued production of primary product
3	Ancillary (support) system to main production process
2	Stand-by unit in a critical system
1	Other ancillary assets

- Work Impact

5	Immediate threat to safety of people and/or plant
4	Limiting operations ability to meet its primary goals
3	Creating hazardous situations for people or machinery, although not an immediate threat
2	Will affect operations after some time, not immediately
1	Improve the efficiency of the operation process

Backlog Management

- Work prioritization also allows making sense out of their maintenance backlog
- Maintenance backlog is very simply
  - The essential maintenance tasks to repair or prevent equipment failures
  - That have not been completed yet
- Maintenance backlogs can be developed.
  - From an overall organizational perspective
  - Within smaller organizational groups
  - Or within categories (i.e., PM, CM)

What Is Maintenance Planning?

Planning = What & How

What should be done? Understanding the definition of the work

- Scope of work may require talking to requester and/or visiting job site
- Better able to identify tools, steps, procedures, specifications, etc.
- May break job into sub-tasks if job too large or complicated

70 CHAPTER 4

Planning = What & How

- What resources/skills are required?
- What materials/tools/equipment are required?
- What safety & environmental concerns should be considered?
- How it will be done & how long it will take? Sequence of work

Why Plan?

- To identify & prepare craft with tools and resources to accomplish timely/efficiently
- Time needed to plan a job properly can be considerable, but it has a high rate of return.
  - Documented by Doc Palmer (Maintenance Planning and Scheduling Handbook) that proper planning can save 1–3 times the resources needed for job execution.
  - If maintenance job repeatable (as most are), then essential to plan work properly because it will have a much higher rate of return.

Planned Job vs. Unplanned Job

| P | W | P | W | P | W |

On-the-job (on – the – run) Planning

| P | W |

A Planned Job

[P] Planning activity   [W] Work activity

WORK MANAGEMENT: PLANNING AND SCHEDULING

The Value of a Planner

	Shop A	Shop B
Wrench time	30%	55%
Planners	0 planner	2 planner / scheduler
Maintenance people	20	18
	40 hr/wk x 30% x 20 = 240 man-hours/week	40 hr/wk x 55% x 18 = 396 man-hours/week
Productive work	240/20 = 12 hrs/wk/person	396/20 = 19.8 hrs/wk/person

Additional productive work gained = (19.8 - 12) / 12 = 65 %

65% more maintenance work completed with same number of people

Symptoms of Ineffective Planning

- Production downtime more than estimated
- Maintenance people standing around waiting
  - For materials or tools
  - For asset/system to be shut down for maintenance work
- Poor work performance
- High rework
- Frequent trips to storeroom
- High stock-out in the storeroom
- Planners being used to expedite parts

Exercise – Developing a Job Plan

- Using Figures 4.7 & 4.8 as guides, develop an actual job plan for one of the following examples or one from the instructor's choosing:
  - Changing the oil in your car
  - Changing a tire on your car
  - Replacing an air filter for your HVAC
  - Mowing your yard

Interdependency Between Planning & Scheduling

- Planning defines what work will be accomplished and how.
- Scheduling identifies when the work will be completed and who will do it.
- Planning & Scheduling are dependent on one another to be effective.

What Is Maintenance Scheduling?

Scheduling = Who and When

- Who will do it? Which work crew and which individual function:
    - May be completed by craft supervisor
    - May be completed daily, weekly, or monthly
- When it will be done?
    - Do we have the resources - material/ skills/tools/ available?
    - Do we have work permits/ clearances?
    - Have we scheduled with least impact to normal operations (integrated schedule)?

WORK MANAGEMENT: PLANNING AND SCHEDULING

Weekly Schedule

	Available Hours				Mech: 40			Elec: 40	
	Scheduled Hours				Mech: 32			Elec: 36	
	Remaining Hours				Mech: 8			Elec: 4	
					Day of Week				
WO	Task	Skill	Est. Hrs	M	T	W	T	F	S	S
12344	Change filters on Hydraulic Systems 16 thru 25	Mech	8	8						
12345	Replace pump 1B for Fuel System X	Mech	24			8	8	8		
12346	Replace pump 1B for Fuel System Y	Elec	24			8	8	8		
12356	Troubleshoot trips offline for Facility Control System (coordinate with facility control system engineer)	Elec	12	8	4					

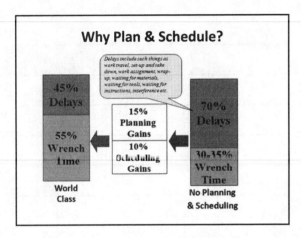

Why Plan & Schedule?

- 45% Delays / 55% Wrench Time — **World Class**
- 15% Planning Gains
- 10% Scheduling Gains
- 70% Delays / 30–35% Wrench Time — **No Planning & Scheduling**

Delays include such things as work travel, set-up and take down, work assignment, wrap-up, waiting for materials, waiting for tools, waiting for instructions, interference etc.

Integrated Schedule

- Rolling annual, quarterly, monthly, or even weekly schedule
  - Includes all types of work
  - Operations activities
  - Maintenance work
  - Capital projects
- Goal: To minimize the impact to any one of these individually but make the best decision for the benefit of the whole facility

74   CHAPTER 4

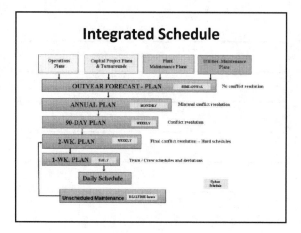

What Is Meant By A Turnaround or Shutdown?

Turnarounds and Shutdowns

- Major downtimes that "just happen" can be disastrous for a plant.
- Planned shutdowns can provide maintenance organizations opportunity
  - Identify and address major potential problems in timely manner to improve safety and efficiency.
  - Includes all maintenance (PM, CM, overhauls, projects) and capital investment projects
  - Typically asset shut down until requested work completed and then restarted, thus "turning around" the asset/facility.

WORK MANAGEMENT: PLANNING AND SCHEDULING

Capital Projects vs. Turnarounds

	Capital Project	Turnaround
Scope	Well defined & static drawings available	Loosely defined, dynamic - changes as inspections made
Planning & Scheduling	Can be planned and scheduled well in advance	Planning & scheduling can't be finalized until scope is approved
Safety permits	Fixed, weekly, or monthly basis	Requires shift and daily basis due to scope functions
Manpower staffing	Fixed, usually doesn't change much	Variable, changes a lot during execution due to scope fluctuations
Schedule update	Weekly or bi-monthly	Shift and daily basis

Capital Projects typically have a project manager while Turnarounds also benefit from similar (a Turnaround Manager)

Work Management / Planning & Scheduling
Measures of Performance

- Percentage of planned work
- Percentage of schedule compliance
- Percentage of time kits delivered on time
- Percentage of time correct parts delivered
- Percentage of work ID from formal PM/CBM
- Percent Rework
- Backlog

Summary

- Effective work management involves a standardized work process with specific roles and responsibilities identified
- Effective work classification and prioritization are important steps
    - In getting the "right" work accomplished
    - Which help properly manage your work backlog
- Planning defines what and how
- Scheduling defines when and who
- Turnaround management can help get work done more efficiently

4.10 Chapter Self-Assessment Questions

1. Draw a workflow chart to show work from a request to completion.
2. Explain each role as shown in the workflow chart from Q 4.1.
3. What is the purpose of a job priority system?
4. Why do we need to manage maintenance backlog? What is a good benchmark?
5. What are the symptoms of ineffective planning?
6. Should planners help schedulers or craft supervisors during an emergency? If yes, explain.
7. Who are key players in scheduling process? Explain their roles.
8. What are the key differences between planning and scheduling processes?
9. Discuss work types and the benefits of work classifications.
10. What are the key differences between capital projects and turnarounds?

WORK MANAGEMENT: PLANNING AND SCHEDULING

Chapter 4 Comprehension Assessment Questions

1. Why is it important for maintenance tasks to be performed efficiently?
2. What is meant by maintenance tasks are executed efficiently and effectively?
3. Give an example whereby a maintenance task was performed using bad and good work management methods.
4. What is "wrench time"?
5. Give some examples of activities that would not be considered "wrench time."
6. What was Abraham Lincoln quoted regarding effective work management practices?
7. How can maintenance craft be more productive then what's sometimes thought of as the typical repair man image (one who just sits or stands around waiting for something to break)?
8. What are the different steps required for effective work management?
9. What percentage of planning should be done for maintenance craft in order to move from reactive to proactive maintenance?
10. What compliance to a work schedule should be sought after for maintenance craft in order to move from reactive to proactive maintenance?
11. What work categories work or classifications of work in the workflow process allow you to skip planning and scheduling and go straight to work execution?
12. What work categories work or classifications of work in the workflow process allow you to skip the planning process and go directly to the scheduling process?
13. What are some potential work order status codes that can be used in routing work through the workflow process?
14. Which work role is responsible for planning the job and creating a work plan or job package that consists of what work needs to be done?
15. Which work role is responsible for working with the craft supervisor and other support staff to develop the weekly, monthly, annual long-range plans to execute maintenance work?
16. Which work role is responsible for taking the weekly schedule and assigning who will do the job on a daily basis?
17. Describe what is meant by a calendar-based PM
18. What type of operating schedule for an asset is a good reason for performing calendar-based PMs?
19. Describe what is meant by a run-based maintenance PM.
20. Describe what is meant by a condition-based PM.
21. Describe what is meant by an operator-based PM.
22. Describe what is meant by a scheduled CM.
23. Describe what is meant by major repairs and projects type of CM.
24. What is the purpose of establishing work priority codes?
25. What are the drawbacks of not clearly defining work parties?
26. What is the benefit of a disciplined method of work prioritization?
27. What work role should set the original work priority for work orders?
28. What work role should validate the work priority?
29. Give an example of an asset criticality set.

78 CHAPTER 4

30. Give an example of work impact priorities.
31. Rank the following work orders based on asset criticality and work impact data.
32. How should one try to make sense out of their maintenance backlog?
33. What is one potential negative impact for an organization that moves toward proactive maintenance?
34. What is a simple definition of planning?
35. What is a simple definition of scheduling?
36. What is the ultimate goal of the planning process?
37. What types of maintenance jobs produce an higher rate of return than would be gained for planning regular jobs?
38. Compare two maintenance shops by calculating their productivity based on the number of maintenance craft and planners that they have, choosing which maintenance shop gets more work done.
    Shop A: 18 maintenance craft, 0 planner/schedulers or Shop B: 12 maintenance craft, 1 planner/scheduler
39. What skills and experience should a person performing the planner/scheduler function have?
40. What must the planner do if the scope of the work has not been defined clearly?
41. What can be used to estimate jobs?
42. What are essential to good job estimating practices?
43. What skills or previous experiences would be good for individuals who become a planner?
44. What should be included within a job plan?
45. What are some techniques for enhancing planning capability?
46. What does scheduling ensure?
47. How is scheduling a joint maintenance and operations activity?
48. Develop a job plan for a specific maintenance activity of your choosing.
49. Develop a weekly schedule for a 40-hour week based on the following information for a specific work crew.
    This work crew includes 1 Electrical Technician and 1 Mechanical Electrician for 40 hours/week. This work crew also shares 1 Laborer with another crew so it uses this individual 20 hours/week. Work orders prioritized to be worked this week are listed below:

12345	Clean oil tank AB (1 Laborer 4 hours)
78320	Perform infrared scans of electrical yard for facility (1 Elec Tech 4 hours)
57353	Repair motor 3 (1 Elec Tech 16 hours, 1 Mech Tech 16 hours)
59330	Perform 72 mo PM on pump 1 (1 Elec Tech 8 hours for disconnect & re-install total, 1 Mech Tech 24 hours, 1 extra pair hands 16 hours)
93740	Electrical cabinet fab (1 Elec Tech 12 hours)

50. What are the key things a maintenance person will need for the job when the scheduled time arrives for that job?
51. List the six basic scheduling principles identified by the noted authority in the area of maintenance planning and scheduling (Doc Palmer).

52. According to noted authority in the area maintenance planning and scheduling, Doc Palmer, what is the primary measure of work efficiency and planning and scheduling effectiveness?
53. What is the measure of scheduling effectiveness?
54. What is the benefit of a planned shutdown for maintenance organizations?
55. What are some examples of work that can be completed during a turnaround or shutdown?
56. What types of work do turnarounds typically consists of?
57. What is the purpose of identifying and appointing a turnaround planner?
58. What are some items for which attention may be required and appropriate corrective actions planned for a turnaround or shutdown?
59. What is an area in which turnaround planning is most often underestimated?
60. What are some key measures of performance for the work management, planning, and scheduling processes?

Research Questions

(Always cite your references when answering research questions.)

1. Describe the trend in maintenance productivity in the last 50 years using resources outside this book.
2. Research new alternative maintenance work flow processes.
3. Research work order status codes and how they fit into maintenance work flow process.
4. Research work classifications and how they fit into the maintenance work flow process.
5. Describe an alternative work priority system; evaluate whether it is better or worse than the one described in this book and explain why.
6. Research alternative backlog management methods and discuss which is best and why.
7. Research planning best practices other than those described by Doc Palmer.
8. Research scheduling best practices other than those described by Doc Palmer.
9. Research best practices regarding turnarounds/shutdowns.
10. Describe the process whereby wrench time information can be collected in an organization.
11. Discuss improvements to maintenance processes that would increase wrench time.
12. Identify and describe alternative measures of performance related to work management, planning, and scheduling.

Chapter 5: Materials, Parts, and Inventory Management

*Almost all quality improvements comes via simplification
Of design, material, manufacturing layout,
processes, and procedures.*
-- Tom Peters

Chapter 5: Materials, Parts, and Inventory Management

- 5.1 Introduction
- 5.2 Key Terms and Definitions
- 5.3 Types of Inventory
- 5.4 Physical Layout and Storage Equipment
- 5.5 Optimizing Tools and Techniques
- 5.6 Measures of Performance
- 5.7 Summary
- 5.8 Chapter Assessment
- 5.9 References and Suggested Reading

Chapter 5: Objective

- Understand maintenance store operations
- Identify different types of inventory
- Be able to use tools and techniques to optimize inventory
- Know how to ensure availability of parts and materials on time
- Ensure you have effective store room layout and storage equipment

MATERIALS, PARTS, AND INVENTORY MANAGEMENT

Introduction

- Logistics is a part of maintenance often times overlooked
- Maintenance storerooms
- Inventory management
- Parts and materials availability
- Storage equipment

Introduction – Best Practices

Application of best practices related to maintenance logistics has produced reductions of:

- 20% in maintenance planner workload
- 30% in # PO's for replenishment of parts
- 40% in manually prepared direct purchase requests
- 30% in maintenance store inventories
- 20% in total maintenance costs

Introduction – Storeroom Objectives

- Objectives of maintenance storerooms
  - Right spares/parts/material
  - Of the right quantity
  - At the right location or site
  - At the right time
- Maintenance storerooms play key role in support of maintenance function

CHAPTER 5

Introduction – So Why Do We Care?

- If right part not available when needed, repairs will be delayed
- Delays in restoring a failed asset will increase maintenance and operations costs
- Additionally, maintenance technicians spend as much as 20% to 30% of their time looking for the right part

Introduction – When Should We Care?

- Obviously when planning job
- Even before that – at design phase of the asset's lifecycle
    - The right time to decide what parts and material should be stocked (and in what quantity) is before placing assets in service
    - Recommended spare parts identification with design FMEA

What Exactly is Inventory?

What Do Types of Inventory Mean?

What Is Inventory?

- Inventory is the quantity of goods and material on hand (available)
  - Goods for sale
  - Material to use
- Production Inventory Categories
  - Finished goods
  - Work-in-process (WIP)
  - Raw materials
  - Other supplies and consumables

What about Maintenance Inventory?

- Maintenance inventory is a hedge against the unknown
  - Protects against uncertainty of need
  - Protects against uncertainty of delivery
- Maintenance Inventory Categories
  - Active inventory
  - Infrequently-used inventory
  - Rarely-used inventory

Inventory Analysis

Inventory analysis (also known as inventory stratification) is a technique to classify and optimize inventory levels:

- Based on items' value and usage rate
- Distinguish between trivial many and vital few
- Reflects Pareto principle
- Criteria dependent on organization

ABC Inventory Analysis

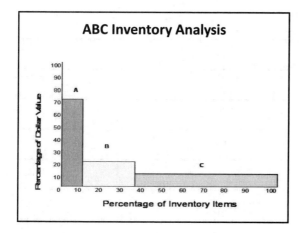

ABC Inventory Analysis Example

Part	Part Description	ABC	Unit Value	Unit Qty	Total Value
1	Part 1 Description		$1,164	49	$57,036
2	Part 2 Description		$7,737	13	$100,581
3	Part 3 Description		$6,742	5	$33,710
4	Part 4 Description		$8,743	9	$78,687
5	Part 5 Description		$9,189	20	$183,780
6	Part 6 Description		$3,020	29	$87,580
7	Part 7 Description		$9,753	11	$107,283
8	Part 8 Description		$1,074	25	$26,850
9	Part 9 Description		$7,700	15	$115,500
10	Part 10 Description		$1,052	26	$27,352
11	Part 11 Description		$2,123	41	$87,043
12	Part 12 Description		$9,705	22	$213,510
13	Part 13 Description		$3,277	37	$121,249
14	Part 14 Description		$4,459	26	$115,934
15	Part 15 Description		$3,766	39	$146,874
					$1,502,969

Watch Out for Hidden Inventory!

- Another often overlooked type of inventory
- Hidden stock (inventory) are items that mechanics stash (under stairwells, within cabinets, or inside toolboxes
- This inventory, thought to be lost, is reordered (at an additional cost)
- Most important, a symptom of a reactive maintenance culture

MAINTENANCE AND RELIABILITY BEST PRACTICES WORKBOOK 87

If It's Already Dead, Then Kill It!

- Other items being stored cannot be classified because it's "dead stock"
  - Spares and materials for assets long removed from operations
- Should be sent back to supplier, sold to surplus or scrap, or maybe even trashed
- Takes up space to store and probably inventoried regularly (additional cost)

Why does it matter how maintenance storerooms are organized
or
where they're located?

Why should I care about storage equipment?

I'm a maintenance organization!

Maintenance Storerooms

- Physical maintenance storeroom attributes are important factor in gaining productivity
- Two considerations for these:
  - Location – Maintenance storerooms should be as close as possible to where maintenance work is performed
  - Layout – Maintenance storerooms should allow for efficient material flow and easy/logical retrieval

88 CHAPTER 5

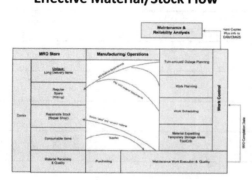

Material Storage and Retrieval

- Many different storage and retrieval methods to handle parts and materials in a storeroom
- Right choice depends on attributes of parts and its demand
- Two major categories of equipment:
  - Man-to-Part
  - Part-to-Man

Storage and Retrieval Equipment

Man-to-Part	Part-to-Man
Shelves/Bins	Horizontal Carousels
Pallet Racks	Vertical Carousels
Modular Drawers	Automated Storage & Retrieval System

Automated Storage and Retrieval System (ASRS)

How Can I Possibly Optimize This Whole Maintenance Logistics Process?

Optimizing Tools and Techniques

- CMMS/EAM data
- "Shelf Life" management
- Inventory accuracy
- Parts kitting
- Economic Order Quantity
- New technologies

CMMS/EAM Data

- Most CMMS/EAM have built-in inventory management system
    - Can be assigned locations
    - Can track usage (purchases/issues)
    - Can associate with correct asset
    - Can be used for inventory classification and analysis

MATERIALS, PARTS, AND INVENTORY MANAGEMENT

"Shelf Life" Management

- Some inventory items, especially spares, require periodic maintenance or checks (lubrication, rotation, etc.)
- Position or orientation of a stored item can damage some parts or reduce life (belts hanging versus flat)
- Some items have limited shelf life (rubbers, chemicals)

Parts Kitting

- One function that storeroom/logistics personnel can provide is parts kitting
- Parts kitting assembles parts, materials, tools, and consumables into groups to enable maintenance technicians in performing planned jobs
- Storeroom personnel can build PM or repair part kits in advance
- This will minimize potential for technician to have to locate parts later

Inventory Accuracy

- Achieving a high level of inventory accuracy is a critical factor for successful storeroom operations.
- Accurate inventory equals actual quantity and types of inventory in the right location according to CMMS/EAM.
- There is a cost for inventory inaccuracy.

Types of Inventory Cost

- Carrying costs: costs of holding items in the storeroom
- Ordering costs: costs of replenishing inventory
- Stock out costs: loss of production when asset cannot be repaired due to stock out or unavailability of parts and materials

Economic Order Quantity (EOQ)

- EOQ analysis is a commonly used technique to optimize inventory levels
- Helps in ordering "right" quantity at specified time interval to meet customer needs while minimizing inventory costs
- Not applicable to every organization, but some may benefit from this type of analysis

Calculating EOQ - Equations

- EOQ (Q) =

$$Q = \sqrt{\frac{2DS}{H}}$$

Where
D = Demand / Usage in units per year
S = Ordering cost per order
H = Inventory carrying cost per unit per year

- Total Annual Inventory Cost =

$$\frac{Q}{2}H + \frac{D}{Q}S$$

EOQ Example

- A plant has a usage rate of 132 oil drums per year.
  - What would be an optimal order quantity?
  - How many orders per year will be required?
  - What would be the additional total cost of ordering and holding these drums in storage if ordering cost is increased by $10/order?
- Plant data also indicates that:
  a) Preparing and receiving oil costs $60/order
  b) The oil drum carrying (holding) cost is 22% per year.
  c) Average cost of an oil drum is $500.

EOQ Example (Cont'd)

$$EOQ\ (Q) = \sqrt{\frac{2 \times 132 \times 60}{110}} = 12$$

$$\text{\# Orders/Year} = \frac{132}{12} = 11$$

$$\text{Total Cost (TC)} = \frac{12}{2}110 + \frac{132}{12}60 = 1320$$

EOQ Example (Cont'd)

$$EOQ\ (Q) = \sqrt{\frac{2 \times 132 \times 70}{110}} \cong 13$$

$$\text{\# Orders/Year} = \frac{132}{13} = 10.2$$

$$\text{Total Cost (TC)} = \frac{13}{2}110 + \frac{132}{13}70 = 1425$$

New Technologies

- New technologies which have been used in supermarkets and transportation companies are finding their way into maintenance logistics
- Automated ID technology:
  - Barcode
  - Radio Frequency Identification (RFID)

How would you best measure success in the area of maintenance logistics?

Measures of Performance

1. % Inactive Inventory
2. % Classification (ABC)
3. Inventory Variance (Inaccuracy)
4. Service Level
5. % Inventory Cost to Plant Value
6. Inventory Shrinkage Rate
7. % Vendor Managed Inventory (VMI)
8. Inventory Growth Rate
9. % Stock Outs
10. Inventory Turnover Ratio

MATERIALS, PARTS, AND INVENTORY MANAGEMENT

Summary

Maintenance logistics is an important part of the maintenance program:
- Main storerooms
- Inventory management
- Parts and material availability
- Storage equipment

Summary (cont'd)

There are multiple objectives related to maintenance logistics:
- Right spares/parts/material
- Of the right quantity
- At the right location or site
- At the right time

Summary (cont'd)

And there are multiple tools and techniques to optimize logistics:
- CMMS/EAM data
- "Shelf Life" management
- Inventory accuracy
- Parts kitting
- Economic Order Quantity
- New technologies

Chapter 5 Self-Assessment Questions

1. Inventories in a plant are generally classified into what categories?
2. What is meant by ABC classification as it is related to inventory?
3. Discuss how the cost of inventory can be optimized.
4. How will you organize a storeroom? Discuss the key features of a small storeroom you have been asked to design.
5. Why is inventory accuracy important? What will you do to improve it?
6. What are the key factors used in calculating EOQ?
7. Explain the benefits of using RFID technology to label stock items–material.
8. Explain inventory turnover ratio. What are the benefits of tracking this ratio?
9. Identify three key performance measures that can be used to manage MRO storerooms effectively.
10. What is meant by shelf life? What should be done to improve it?

MATERIALS, PARTS, AND INVENTORY MANAGEMENT

Chapter 5 Comprehension Assessment Questions

1. What are the objectives of maintenance storerooms?
2. What are two impacts to the right part not being available when needed?
3. What is a common percentage of time that maintenance technicians spend looking for the right parts?
4. What are five benefits to the application of effective maintenance logistics (materials, parts, and inventory management) techniques?
5. What are the two basic functions that storerooms provide?
6. When is the best time to decide what parts and materials should be stocked and in what quantities?
7. What is one analysis technique that is useful in optimizing the spares list, including providing a good estimate of what and how many spares should be stocked during a specific period?
8. What percentage of item cost per year is associated with stocking an item and then holding it in inventory?
9. List and define the three major categories of maintenance inventory based on their usage.
10. What have some organizations done with other organizations to help with the expense of rarely-used inventory items?
11. What is meant by the term inventory stratification?
12. Based on the following data, perform an inventory stratification analysis.

Part	Part Description	ABC	Unit Value	Unit Qty	Total Value	Part	Part Description	ABC	Unit Value	Unit Qty	Total Value
1	Part 1 Description		$2,760	33	$91,080	1	Part 16 Description		$5,885	18	$105,930
2	Part 2 Description		$1,182	44	$52,008	2	Part 17 Description		$2,025	35	$70,875
3	Part 3 Description		$1,363	47	$64,061	3	Part 18 Description		$2,107	46	$96,922
4	Part 4 Description		$4,863	32	$155,616	4	Part 19 Description		$455	54	$24,570
5	Part 5 Description		$3,896	36	$140,256	5	Part 20 Description		$6,563	12	$78,756
6	Part 6 Description		$1,056	31	$32,736	6	Part 21 Description		$4,437	49	$217,413
7	Part 7 Description		$6,259	10	$62,590	7	Part 22 Description		$2,037	43	$87,591
8	Part 8 Description		$2,687	34	$91,358	8	Part 23 Description		$2,457	46	$113,022
9	Part 9 Description		$3,297	45	$148,365	9	Part 24 Description		$1,342	41	$55,022
10	Part 10 Description		$7,885	17	$134,045	10	Part 25 Description		$5,600	23	$128,800
11	Part 11 Description		$8,489	20	$169,780	11	Part 26 Description		$2,422	29	$70,238
12	Part 12 Description		$8,488	13	$110,344	12	Part 27 Description		$5,972	5	$29,860
13	Part 13 Description		$5,948	7	$41,636	13	Part 28 Description		$6,023	7	$42,161
14	Part 14 Description		$5,612	4	$22,448	14	Part 29 Description		$3,292	41	$134,972
15	Part 15 Description		$8,133	24	$195,192	15	Part 30 Description		$3,636	29	$105,444
											$2,873,091

13. What is meant by the term hidden stock inventory?
14. What is a potential problem that comes from hidden stock inventory?
15. What is meant by the term dead stock?
16. What are two issues involved in decisions about the physical layout of the maintenance store?
17. Where should a maintenance storeroom be located?
18. What are the two main categories of part storage equipment and examples from each of these categories?

19. What are some examples of preventive maintenance actions that may be performed on certain parts equipment that are stored?
20. What are some examples of parts that have limited shelf life?
21. What is the preferred percentage value of inventory accuracy?
22. What is meant by the term parts kitting?
23. Define the 3 different components of inventory costs.
24. What is meant by economic order quantity (EOQ)?
25. Calculate economic order quantity, total inventory cost, and potential cost increase based upon the following information.

    A plant has a usage rate of 360 hydraulic filters per year for the entire operations facility. Ordering costs are $8 per order, carrying costs are 20% per year, and average filter cost is $20. There is likely to be an increase ordering cost of $100 per order next year.
26. Draw the economic order quantity and stocking levels diagram.
27. What are some of the new technologies that apply to materials, parts, and inventory management?
28. What are some of the benefits gained from the new technologies that apply to materials, parts, and inventory management?
29. What is the major difference between RFID and barcode technology?
30. What are some advantages that RFID provides over traditional barcoding?
31. What are the best practices related to improving materials, parts, and inventory management and storeroom effectiveness?

MATERIALS, PARTS, AND INVENTORY MANAGEMENT

Research Questions

(Always cite your references when answering research questions. All research questions should include resources outside this book.)

1. Discuss the amount of time that maintenance technicians spend locating parts and materials.
2. Discuss the percentage of the overall maintenance budget dedicated to spare parts.
3. Research the benefits of applying best practices related to materials, parts, and inventory management.
4. Explain why you do or do not think that the best time to plan for the right material and material quantities is in the design phase.
5. Discuss the percentage of inventory cost needed for stocking and holding an item in inventory.
6. Compare different classification methods for maintenance inventory.
7. Compare different techniques used to classify and optimize inventory levels.
8. Evaluate techniques used to manage and organize maintenance storerooms.
9. Describe best practices related to well-organized maintenance storerooms.
10. Evaluate new trends in inventory storage and retrieval equipment.
11. Describe the negative consequences of inventory inaccuracy.
12. Explain processes used to perform parts kitting.
13. Evaluate at least one technique other than economic order quantity that can be used to optimize inventory levels.
14. Describe new trends in technologies used to better manage inventories.
15. Research measures of performance that could be used to assess inventory management not included in this book.

Chapter 6
Measuring and Designing for Reliability and Maintainability

Insanity is doing the same thing, over and over again and expecting different results

- Albert Einstein

Chapter 6: Measuring and Designing for Reliability and Maintainability

6.1 Introduction
6.2 key Terms and Definitions
6.3 Defining and Measuring Reliability and Other Terms
6.4 Designing and Building for Maintenance and Reliability
6.5 Summary
6.6 Self Assessment Questions
6.7 References and Suggested Reading

Chapter 6 Objectives

- Understand reliability and why it is important
- Calculate reliability, availability, and maintainability
- Measure and specify reliability
- Review design for reliability
- Explain the impact of O&M costs on an asset's life cycle cost

MEASURING AND DESIGNING FOR RELIABILITY AND MAINTAINABILITY

Introduction

- Asset reliability important focal point for many organizations
- Source of competitive advantage
- Central theme for maintenance departments
- Identifies the right work to be performed

Reliability Only One Piece

Reliability is design attribute considered when designed and installed

- Asset
- Designed-In Reliability
- Maintenance Plan
- Operating
- Environment

Reliability Also Has 3 Key Elements

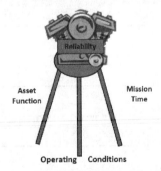

- Reliability
- Asset Function
- Mission Time
- Operating
- Conditions

Maximizing Reliability

- One challenge in understanding reliability is belief of some that they need to absolutely maximize reliability
- 100% reliability often results in high acquisition and maintenance costs not a cost-effective, affordable strategy
- Best approach is to define asset or facility reliability requirements to support underlying business need

Optimizing Reliability

Graph showing Cost vs. Reliability (Availability) with curves for Total Cost, Reliability, and Production & Use Costs, indicating an Optimum Level.

Why Reliability?

Reliability is important for several reasons:
- Customer satisfaction
- Reputation
- Operations & Maintenance (O&M) costs
- Repeat business
- Competitive advantage

MEASURING AND DESIGNING FOR RELIABILITY AND MAINTAINABILITY

The Costs of "Unreliability"

- There are costs of assets not being reliable ("unreliability")
- Loss of production or service not provided
- Lost customer confidence
  - May not buy product or service again
  - Inconvenience effect
  - Time wasted
- Cost to repair or replace components or entire asset

What Is RAM?

Reliability
Availability
Maintainability

Reliability Is Not RCM

For many, reliability is synonymous with Reliability-Centered Maintenance (RCM).
- But reliability is not RCM.
- RCM is a proactive methodology utilizing reliability principles for identifying the right work.
- RCM sustains reliability.

Reliability Is Not Maintenance

For others, reliability is synonymous with maintenance
- But reliability is not maintenance
- Maintenance is the act of maintaining or the actual work of keeping an asset in proper operating condition
- Maintenance also sustains reliability

Reliability Is Not Quality Control

For still others, reliability is synonymous with Quality Control (QC). But reliability is not QC.
- QC is concerned with how process is meeting specifications to guarantee consistent product quality.
- Thus QC is more of a snapshot of manufacturing process quality rather than reliability.
- QC works with reliability.

What Is Reliability?

- Reliability is a broad term focusing on an asset's ability to perform its intended function to support manufacturing, provide a service, or some other capability.
- MIL-STD-721C: "the probability that an item will perform its intended function for a specified interval understated conditions."
- Understanding the term and how it differs from maintenance is key to establishing a successful program for improving reliability for the long-term.

MEASURING AND DESIGNING FOR RELIABILITY AND MAINTAINABILITY

Inherent Reliability

- Assets are designed with a certain level of reliability based on effective use of reliable components and their configurations.
- Some components may work in series and others in parallel to provide the overall desired reliability.
- This is the inherent reliability.

Two Types of Assets

- Repairable assets (or components)
  - Can be repaired when they fail
  - Examples: compressors, hydraulics, pumps, motors, valves
  - Characterized by Mean Time Between Failures (MTBF)
- Non-repairable assets (or components)
  - Cannot be repaired – must replace when fails
  - Examples: light bulbs, rocket motors, circuit boards
  - Characterized by Mean Time to Failure (MTTF)

Mean Time Between Failures (MTBF)

$$MTBF = \frac{\sum Operating\ time}{\#Failures}$$

- MTBF is inverse of failure rate (λ)
- Higher MTBF is better

Mean Time to Failure (MTTF)

$$MTTF = \frac{\sum TimeToFailure}{\#Failures}$$

- Higher MTTF is better

Trending MTBF (and MTTF)

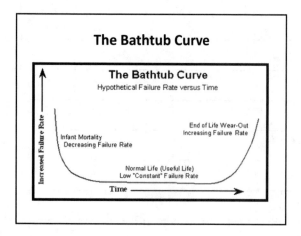

The Bathtub Curve

The Bathtub Curve
Hypothetical Failure Rate versus Time

- Infant Mortality — Decreasing Failure Rate
- Normal Life (Useful Life) — Low "Constant" Failure Rate
- End of Life Wear-Out — Increasing Failure Rate

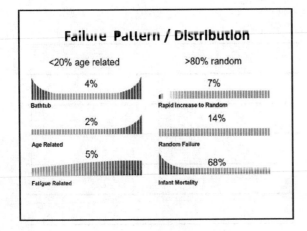

Failure Pattern / Distribution

<20% age related: Bathtub 4%, Age Related 2%, Fatigue Related 5%

>80% random: Rapid Increase to Random 7%, Random Failure 14%, Infant Mortality 68%

Calculating Reliability from MTBF

$$R = e^{-\lambda t}$$

OR

$$R = e^{-t \frac{1}{MTBF}}$$

*Recall that $\lambda = \frac{1}{MTBF}$

Calculating Required MTBF from Desired Reliability

- In order to reverse this equation, we simply take the natural log (ln) of both sides of the previous equation, which eventually translates to the following formula:

$$MTBF = -t \frac{1}{\ln(DesiredReliability)}$$

Maintainability

- Maintainability is a measure of an asset's ability to be restored to a specified condition when maintenance is performed by personnel having certain skills and using prescribed procedures and adequate resources

- Characterized by Mean Time to Repair (MTTR)

Mean Time to Repair (MTTR)

$$MTTR = \frac{\sum Time\,To\,Repair}{\#\,Failures\,(Repairs)}$$

- Some call it pure repair time (or wrench time).
- Others use Mean Downtime (MDT) to capture actual time asset is down.
- Lower MTTR is better.

Trending MTTR

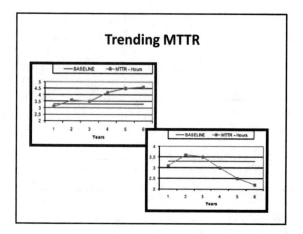

MEASURING AND DESIGNING FOR RELIABILITY AND MAINTAINABILITY

Availability

- Availability is a function of reliability and maintainability.

- It is measured by the degree to which an asset or facility is operating at the start of the mission and for its duration when the asset or facility is needed

Different Types of Availability

- *Inherent Availability:* Classical definition determined by design

- *Achieved Availability:* Includes preventive maintenance but not delays

- *Operational Availability.* Considers preventive and corrective maintenance, including any delays

Availability

$$Availability = \frac{MTBF}{MTBF + MTTR}$$

OR

$$Availability = \frac{Uptime}{Uptime + Downtime}$$

Calculating RAM Example

A plant's air compressor system is operated for 1000 hours last year. The plant's CMMS system provided the following system data:
- Operating time = 1000 hours
- Number of failures, random = 10
- Total hours of repair time = 50 hours

What's the expected reliability and availability of this compressor system if we have to operate this unit for 100 hours?

Calculating RAM Example (cont'd)

$$MTBF = \frac{1000}{10} = 100$$

$$MTTR = \frac{50}{10} = 5$$

Calculating RAM Example (cont'd)

$$R = e^{-(100)\frac{1}{100}} = 0.37 = 37\%$$

$$Availability = \frac{100}{100+5} = 0.95 = 95\%$$

MEASURING AND DESIGNING FOR RELIABILITY AND MAINTAINABILITY

What If?

- What would the MTBF value have to be if the reliability needed was 75% in order to meet customer expectations?

- Based on the previously calculated MTBF value, what would the MTTR value have to be if the availability needed was 99% in order to meet customer expectations?

What If Example (cont'd)

$$MTBF = -(100)\frac{1}{\ln(0.75)}$$

$$MTBF = 347.6$$

- Let's check ourselves:

$$R = e^{-(100)\frac{1}{347.8}} = 0.75 = 75\%$$

What If Example (cont'd)

$$MTTR = \frac{MTBF}{Availability} - MTBF$$

$$MTTR = \frac{347.8}{0.99} - 347.8 = 3.5$$

$$Availability = \frac{347.8}{347.8 + 3.5} = 99\%$$

114 CHAPTER 6

What Is Reliability Modeling?

Reliability Modeling

- Failure logic of assets, components, or even entire facilities can be shown in a Reliability Block Diagram (RBD).

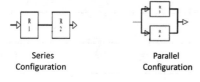

Series Configuration   Parallel Configuration

Reliability Modeling

- Some failure logic in a Reliability Block Diagram (RBD) includes m – out – of –n that need to function.

m-out-of-n configuration

MEASURING AND DESIGNING FOR RELIABILITY AND MAINTAINABILITY

Calculating Series Reliability

$$R_{sys} = R_1 \times R_2 \times \ldots R_n$$

$$R_{sys} = e^{-t(\lambda_1 + \lambda_2 + \ldots + \lambda_n)}$$

Calculating Parallel Reliability

$$R_{sys} = 1 - (1 - R_3)(1 - R_4)\ldots(1 - R_n)$$

OR

$$R_{sys} = 1 - (1 - e^{-t(\lambda_3)})(1 - e^{-t(\lambda_4)})\ldots(1 - e^{-t(\lambda_n)})$$

Calculating m-out-of-n Reliability

m - # of Working element	Overall System Reliability
1 out of 2	$R^2 + 2R(1-R) = 1 - (1-R)^2$
2 out of 2	R^2
1 out of 3	$R^3 + 3R^2(1-R) + 3R(1-R)^2 = 1 - (1-R)^3$
2 out of 3	$R^3 + 3R^2(1-R)$
3 out of 3	R^3
1 out of 4	$R^4 + 4R^3(1-R) + 6R^2(1-R)^2 + 4R(1-R)^3 = 1 - (1-R)^4$
2 out of 4	$R^4 + 4R^3(1-R) + 6R^2(1-R)^2$
3 out of 4	$R^4 + 4R^3(1-R)$
4 out of 4	R^4

Assumption: each element/component's reliability is same = R

Calculating System Reliability

- The real world isn't as simple as one series or parallel or m-out-of-n configuration.
- The real world is usually a combination of series, parallel, & m-out-of-n configurations.
- Therefore, the key to calculating system reliability with various configuration types is to first break the complex into simple.
  - Convert every parallel and m-out-of-n configuration into a single reliability value.
  - Then each of these single reliability values can be treated as a series group together.

Calculating System Reliability Example

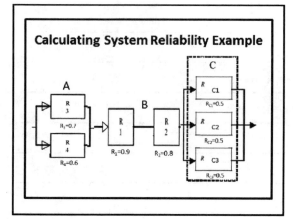

Calculating System Reliability Example

$$R_A = 1 - (1 - 0.7)(1 - 0.6) = 0.88$$

$$R_B = (0.9)(0.8) = 0.72$$

$$R_C = (0.5)^3 + 3(0.5)^2(1 - 0.5) = 0.5$$

$$R_{ABC} = (0.94)(0.72)(0.5) = 0.32$$

MEASURING AND DESIGNING FOR RELIABILITY AND MAINTAINABILITY

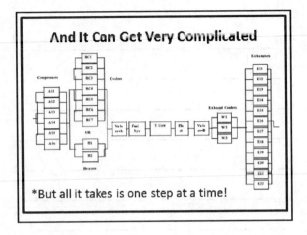

And It Can Get Very Complicated

*But all it takes is one step at a time!

Reliability Modeling

- Then it can get too complicated to perform these calculations by hand.
- There are multiple software packages available to perform this modeling.
    - Input MTBF & MTTR values for each component.
    - Typically will assume exponential failure distribution.
    - But one can even input actual failure data if software package will create failure distributions (Weibull, etc.).

What Is Meant by Asset Life Cycle Cost (LCC)?

Asset Life Cycle Cost (LCC)

- Asset LCC are ALL costs expected during the life of the asset.

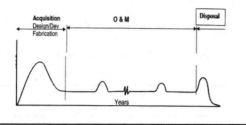

Asset Life Cycle Cost (LCC)

- ALL costs throughout the life of asset
- Acquisition
  - Design and Development
  - Demonstration and & Validation
  - Manufacturing/Assembly/Build/Installation
- Operations & Maintenance (O&M)
  - Operations (includes manpower, energy, supplies, etc.)
  - Maintenance (PM, CM, etc.)
- Disposal

Distribution of Asset Life Cycle Cost (LCC)

For a Typical	DoD System*	Industrial
Design and Development	10–20%	5–10%
Production / Fabrication / Installation	20–30%	10–20%
Operations and Maintenance (O&M)	50–70%	65–85%
Disposal	<5%	<5%

* Assurance Technologies Principles and Practices by Raheja & Allocco DoD –Department of Defense

MEASURING AND DESIGNING FOR RELIABILITY AND MAINTAINABILITY

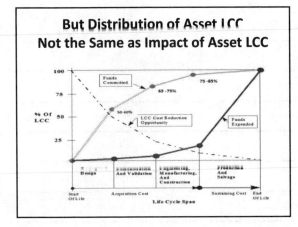

How Does One Design for RAM (Reliability, Availability, and Maintainability)?

Design for RAM

- RAM requirements and specifications
- Proper component selection and configuration
- Reviewing the design based on RAM
- Considering logistics support
- Planning in order to reduce O&M costs

CHAPTER 6

RAM Requirements and Specs

- A reliable asset must have good reliability (RAM) requirements and specifications
- Should address:
  - Condition and environment asset will operate in, probability of successful performance, function to perform, mission time, usage limitations, skills needed for operators & maintainers
- Should be unambiguous since otherwise leaves great deal of room for error, which leads to poorly-understood design requirements

Other Good Ideas for RAM Specs

- Display asset performance/condition data (early warnings, diagnostic display, pinpoint problems)
- Use of modular and standardized components
- Use of redundant parts/components
- Minimize special tools — parts
- Operations and maintenance training material
- FMEA / RCM-based maintenance plan
- Maximizes use of CBM technologies
- Basis for spares recommendations
- Life Cycle Cost analysis
- O&M cost estimates

Good Example of RAM Requirements

There shall be a 90% probability (of success) that the cranking speed is more than 85 rpm after 10 seconds of cranking (mission) at $-20\,^{\circ}F$ of (environment) for a period of 10 years or 100,000 miles (time). The reliability shall be demonstrated at 95% confidence.

Assurance Technologies Principles and Practices by Raheja and Allocco

*Additionally, see checklist on pp. 182-184 in the book

MEASURING AND DESIGNING FOR RELIABILITY AND MAINTAINABILITY

RAM Approach in Design

- As much as 60% failures and safety issues can be prevented by making changes in design
- Asset should be designed:
  - For fault tolerance
  - To fail safely
  - With early warning of failure to user
  - With built-in diagnostics to ID fault location
  - To eliminate all critical fail modes cost-effectively

RAM Analysis for Design

- Reliability Analysis
- Maintainability Analysis
- System Safety and Hazard Analysis
- Human Factors Engineering Analysis
- Logistics Analysis

How Do We Improve RAM?

How Do We Measure These Improvements Against Where We Currently Are?

Improving RAM

- After an asset is installed, reliability cannot be changed without redesign or replacement of better components.
- Maintainability can be improved with changes to training, planning and scheduling, logistics, available spares and materials, etc.
- Availability can be improved by repairing or replacing bad components before they fail with a good reliability-based (RCM) program
- Evaluating and finding ways to improve these is a key aspect of reliability engineering.

Designing for RAM
Measures of Performance

- Reliability (MTBF, MTTF)
- Availability
- Maintainability (MTTR)
- Asset Life Cycle Cost (LCC)
- % Designs with RAM Requirements

Summary

- Improving reliability essential to the success of any organization, especially operations and maintenance groups
- Understanding reliability and its relation to maintenance in its reduction of overall asset life cycle costs (LCC)
    - O&M is ~80% of overall LCC
    - O&M costs are mostly determined during design
- Reliability should be designed-in and should be a strategic element of facility planning

Chapter 6 Self-Assessment Questions

1. Define reliability and maintainability.
2. What's the difference between maintenance and maintainability?
3. If an asset is operating at 70% reliability, what do we need to do to get 90% reliability? Assume assets will be required to operate for 100 hours.
4. If an asset has a failure rate of 0.001 failures/hour, what would be the reliability for 100 hours of operation?
5. What would be the availability of an asset if its failure rate is 0.0001 failures/hour and average repair time is 10 hours?
6. What would be the availability of a plant system if it is up for 100 hours and down for 10 hours?
7. If an asset's MTBF is 1000 hours and MTTR is 10 hours, what would be its availability and reliability for 100 hours of operation?
8. Define availability. What strategies can be used to improve it?
9. What is the impact of O&M cost on the total life cycle cost of an asset?
10. What approaches could we apply during the design phase of an asset to improve its reliability?

CHAPTER 6

Ch 6 Comprehension Assessment Questions / Answers

1. What is the difference between reliability and reliability centered maintenance (RCM)?
2. What is the relationship between RCM and PM?
3. What are three key elements of asset reliability?
4. What is the relationship between reliability and design?
5. What is the objective of maintainability?
6. What is required in order to improve reliability?
7. What is the downside of ensuring 100% reliability?
8. Draw the reliability/availability economics diagram.
9. What are the basic reasons that asset reliability should be considered important?
10. What are the two basic types of assets with regard to measuring and designing for reliability and maintainability?
11. What is the difference between MTTR and MDT?
12. Why would one say that MTBF and MTTR have reverse relationships?
13. What are the differences between inherent availability, operational availability, and achieved availability?
14. Describe what is meant by the bathtub curve.
15. Draw and give percentage of population values for the six basic failure patterns.
16. How much more likely is it to have a random failure than an age-related failure?
17. Calculate reliability and availability values based on the following information.

    A facility's critical piece of equipment is operated for 4500 hours last year. According to available data, this equipment had 15 failures with an average of 40 hours required to repair each failure. The facility manager would like to estimate the expected probability that this equipment can make it through a critical operation for 500 hours without a failure along with the expected availability for next year's operation.
18. Based on the previous data, what would it take to reach the facility manager's desired reliability value of 80%?
19. Assuming the desired reliability value was attained from the previous question, would this alone enable an availability value of 99%?
20. Calculate reliability and availability values based on the following information.

    A facility's emergency generator is operated for 250 hours last year. According to available data, this equipment had 5 failures with an average of 16 hours required to repair each failure. The facility manager would like to estimate the expected probability this equipment can make it through a typical downtime period (72 hours) without a failure along with the expected availability for next year's operation (estimated to be about the same as last year).
21. Based on the previous data, what would it take to reach the facility manager's desired reliability value of 80%?
22. Assuming the desired reliability value was attained from the previous question, would this alone enable an availability value of 95%?
23. Draw a reliability block diagram and calculate the overall system reliability for the following example.

The production line has 5 pieces of equipment that must all work in order to manufacture the product. These 5 pieces of equipment have reliability values of 0.95, 0.98, 0.88, and 0.92 respectively.

24. Using the same data as the previous question, draw a reliability block diagram and calculate the overall system reliability if only 1 out of the 4 pieces of equipment was needed to manufacture the product.
25. Assuming that the reliability of each individual piece of equipment was improved to the highest level of reliability for the single piece of equipment, draw a reliability block diagram and calculate the overall system reliability if 2 of the pieces of equipment were needed to manufacture the product.
26. Describe what is meant by the term asset life cycle cost.
27. What must one focus on in order to design for reliability and maintainability
28. What is the problem with ambiguous reliability specifications?
29. What are the key elements of reliability specifications?
30. Other than the key elements of reliability specifications, what are the other requirements and specifications that should be included?
31. What analyses should be performed during the design phase?
32. Give at least 10 different examples of what should be included on a design review checklist.

Research Questions

(Always cite your references when answering research questions.)

1. Using resources outside this book, discuss the trend in maintenance productivity in the last 50 years.
2. Research new alternative maintenance work flow processes.
3. Research work order status codes and how they fit into maintenance work flow process.
4. Research work classifications and how they fit into the maintenance work flow process.
5. Describe an alternative work priority system. Evaluate whether it is better or worse than the one described in this book and why.
6. Research alternative backlog management methods. Which do you think is the best? Explain why you selected that method.
7. Research planning best practices other than those described by Doc Palmer.
8. Research scheduling best practices other than those described by Doc Palmer.
9. Research best practices regarding turnarounds/shutdowns.
10. Describe the process whereby wrench time information can be collected in an organization.
11. Discuss improvements to maintenance processes that would increase wrench time.
12. Identify and describe alternative measures of performance related to work management, planning, and scheduling.

Chapter 7
Operator Driven Reliability

Reliability cannot be driven by the maintenance organization. It must be driven by the operations...and led from the top.
— Charles Bailey

Chapter 7: Operator Driven Reliability

7.1 Introduction
7.2 Key Terms and Definitions
7.3 The Role of Operations
7.4 Total Productive Maintenance (TPM)
7.5 Workplace Organization: 5S
7.6 Overall equipment Effectiveness (OEE)
7.7 Measures of Performance
7.8 Summary
7.9 Self Assessment Questions
7.10 References and Suggested Reading

Chapter 7's Objective ... to understand

- The role of operators in sustaining and improving reliability
- Total Productive Maintenance (TPM) and its implementation
- Overall Equipment Effectiveness (OEE)
- Workplace design
- Implementation of the 5(6) -S program to optimize productivity

Introduction

- The objective of the maintenance and reliability organization is:
  - To ensure that assets are available to produce quality products
  - To provide quality service in a cost-effective manner when needed
- The performance of an asset is based on three factors:
  - Reliability (inherent) — How was it designed?
  - Maintenance plan — How will it be maintained?
  - Operating environment — In what environment and with what methods will it be operated?

Operating Environment

The third factor, the operating environment, will be discussed in this chapter.

- This factor includes (1) skills of the operators and (2) operating conditions under which the asset performs.
- Several studies indicated > 40% of failures directly result from operational errors or unsuitable operating conditions.
  - Failures due to faulty inherent design can also be minimized or eliminated if operators understand the asset and how its operation affects overall performance.
  - Operators must feel responsible for proper operation of assets under their control because they sense if something is wrong or abnormal about that asset's operation.

Not an Operator-Involved Culture

- Many organizations have not successfully involved operators in maintenance because of previous work cultures that did not encourage operator involvement.
- Two primary reasons for previous work culture:
  - Division of operations and maintenance labor
  - Historical reward system
- Division of labor prevented operator-driven reliability:
  - Production/operations department operates the assets, without regard to maintenance needs or proper operating conditions.
  - Maintenance department fixes assets when they break, restoring assets to an optimum operational state.
- Reward system misaligned with operator-driven reliability:
  - Design rewarded for achieving functional capability at lowest cost.
  - Operations rewarded when meeting or beating production.
  - Maintenance teams rewarded for fixing asset failures.

What Should Be Operations' Role in Maintenance?

Operator-Involved Culture

- Operations and maintenance need to work together cohesively to deliver high-quality products waste-free and cost-effectively.
- Virtually every major management philosophy and methodology today recognizes and fosters the integral relationship between these.
- The Just-In-Time (JIT) and Lean Manufacturing methods would not be possible without high levels of asset reliability and availability, driven by active operator involvement in maintenance.

Operator-Driven Reliability

- Operator involvement is an integral part of an overall proactive maintenance strategy.
- This concept is best characterized as:
  - Operator-driven reliability (ODR)
  - Operator-based reliability (OBR)
- ODR's objective is to help plants run better, longer, cost-effectively, and competitively by:
  - Reducing unplanned downtime
  - Increasing uptime of production processes and associated assets
  - Proactively identifying problems to eliminate or reduce failures, thereby increasing reliability
- This concept is similar to TPM.

What Is TPM?

TPM

- Total Productive Maintenance (TPM):
  - Is a team-based asset management strategy
  - Emphasizes cooperation between operations and maintenance departments
- TPM has multiple objectives and goals:
  - Achieve zero defects, zero breakdowns, and zero accidents within all functional areas of the organization.
  - Involve people at all levels of the organization.
  - Implement an effective workplace design.

TPM Principles

TPM is based around the following principles:
- Improving asset and equipment effectiveness
- Autonomous maintenance by operators
- Servicing, adjustments, and minor repairs
- Planned maintenance by maintenance department
- Training to improve operations and maintenance skills
- Better workplace design, including cleanliness and the standardization of procedures

TPM Pillars

- *Pillar #1* Autonomous Maintenance
- *Pillar #2* Focused Improvement — Kaizen
- *Pillar #3* Planned Maintenance
- *Pillar #4* Quality Maintenance
- *Pillar #5* Training and Development
- *Pillar #6* Design and Early Equipment Management
- *Pillar #7* Office Improvement
- *Pillar #8* Safety, Health, and Environment

TPM Steps

1. Announce TPM.
2. Launch formal education program.
3. Create an organization support structure.
4. Establish basic TPM policies and quantifiable goals.
5. Outline detailed master deployment plan.
6. Kick off TPM.
7. Improve effectiveness of each piece of equipment.
8. Conduct training to improve operation and maintenance skills.
9. Develop an early equipment management program.
10. Continuously improve.

TPM vs. TQM

Total Productive Maintenance (TPM) closely resembles Total Quality Management (TQM):

- Total commitment to the program is needed from upper level management.
- Employees must be empowered to initiate corrective action.
- A long-term strategy is required; it may take a long time to implement programs and make them a part of the routine, ongoing process.
- Cultural change — new mindsets are required.

Benefits of TPM
- Increased productivity
- Reduced manufacturing cost
- Reduction in customer complaints
- Satisfy the customer's needs by 100%
    - Delivering the right quantity
    - At the right time
    - With best, required quality
- Reduced safety incidents and environmental concerns

What Does 5S Mean?

What Is 5S?
- A technique to reduce waste and optimize productivity by maintaining an orderly workplace and using visual cues to achieve more consistent operational results.
- A structured program to achieve total organization-wide cleanliness and standardization in the workplace.
- 5S stands for 5 Japanese words that have been translated into their English equivalents:
    - Sort (Seiri)
    - Set-in-Order (Seiton)
    - Shine (Seiso)
    - Standardize (Seiketsu)
    - Sustain (Shitsuke)

Many organizations have added sixth S-word to represent Safety.

134 CHAPTER 7

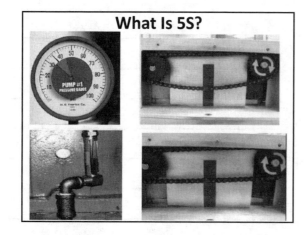

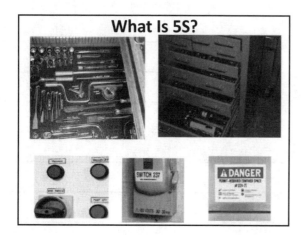

OEE or TEEP?

Both or Neither?

OEE vs. TEEP

- Overall Equipment Effectiveness (OEE)
  - OEE is a key metric used in TPM and Lean Manufacturing programs to measure the effectiveness of TPM and other initiatives.
  - It provides an overall framework for measuring production efficiency.
- Total Effective Equipment Performance (TEEP)
  - TEEP is another related metric that is based on 24 hours per day and 365 days per year operations.
  - TEEP also considers equipment utilization.

OEE vs. TEEP

- Overall Equipment Effectiveness (OEE)
  OEE = Availability × Performance × Quality
- Total Effective Equipment Performance (TEEP)
  TEEP = Utilization (U) × Availability (A) × Performance (P) × Quality (Q)

  or

  TEEP = Utilization (U) × OEE

Calculating OEE

- A given asset, a machining center, experiences the following:

  Asset Availability = 88%
  Asset Performance = 93%
  Production Quality = 95%

  OEE = (A) × (P) × (Q)
  OEE = (88%) × (93%) × (95%)
  OEE = 77.75%

Calculating TEEP

- Using data from the previous example and operation of 20 hours/day, 300 days/yr:

    OEE = 77.75%

    Utilization = (20 × 300) / (24 × 365)
    = 68.49%

    TEEP = (OEE) × (U)
    = 53.25%

    Additional example shown on pages 210-213 in book.

Operator-Driven Reliability Measures of Performance

- Downtime as % Scheduled Hours
- 5S Audit Results
- Asset Performance (Throughput)
- % Ops Personnel Involved in Improvements
- % Operators Qualified/Certified
- % Assets Ready
- OEE
- TEEP

Summary

- Operator-driven reliability (ODR) and Total Productive Maintenance (TPM) strategies encourage participation of all employees, especially operators.
- Under ODR/TPM concept, equipment operators become owners of their assets, working closely with maintainers to preserve assets in best possible condition.
- Operators perform various maintenance:
    - Clean & lubricate equipment
    - Minor adjustments, checks, and fixes

Summary

- 5S (a visual workplace system)
  - Can achieve total organization-wide cleanliness and standardization of the workplace.
  - Results in a safer, more efficient, and more productive operation.
  - Boosts morale of employees, promoting a sense of pride in their work and ownership
- 5S has 5 elements:
  - Sort, Set-in-Order, Shine, Standardize, and Sustain
  - 6th element (Safety) sometimes added

Summary

- Overall Equipment Effectiveness (OEE)
  - Key metric that quantifies how well an asset or manufacturing process performs relative to its designed capacity during periods when it is scheduled to run
  - OEE = Availability × Performance × Quality
- Total Equipment Effectiveness Performance (TEEP):
  - Related to OEE
  - Includes Utilization component
  - TEEP = OEE × Utilization

Chapter 7 Self-Assessment Questions

1. Explain operator-driven reliability. Why is the operator's involvement important in maintenance?
2. Define TPM. What are TPM's various elements (the pillars of TPM)?
3. How do we implement TPM?
4. Define OEE. How do we measure it?
5. What is the difference between OEE and TEEP?
6. Explain 5S. What benefits do we derive from implementing 5S?
7. What is the difference between 5S plus or 6S (and traditional 5S)?
8. Explain what is meant by the visual workplace.
9. Explain Muda, Mura, and Muri.
10. What are the benefits of standardizing?

Chapter 7 Comprehension Assessment Questions

1. What is the objective of a maintenance and reliability organization?
2. What are the three factors that the performance of an asset based upon?
3. Approximately what percentage of failures is the direct result of operational errors or unsuitable operating conditions?
4. What is one result of operators not identifying inciplent problems and correcting them in time?
5. What are the two primary reasons for a work culture that does not involve operations in maintenance activities?
6. What is the objective of OEE?
7. What is utilization rate?
8. What basic maintenance activities outside their normal operator duties may the operator perform under an operator-driven reliability concept?
9. What are the principles on which TPM is based upon?
10. Describe the origin of TPM.
11. What are the similarities between the TPM and TQM programs?
12. What are the similarities when implementing TPM and TQM?
13. What are the objectives of TPM?
14. What are the benefits of TPM?
15. How does TPM create a positive work culture and environment?
16. What are the goals of autonomous maintenance?
17. One of the goals of focused improvement activities such as Kaizen?
18. What are six major losses that can become a focus area of kaizen teams to improve effectiveness of an organization?
19. What factors are taken into consideration for maintenance prevention design?
20. What are the 5 elements of 5S?
21. Draw a diagram to illustrate the concepts of OEE and TEEP and relationship between the two calculations.
22. Calculate OEE and TEEP for a facility based on the following data: 80% availability, 99% performance, 95% quality, and 75% utilization.
23. Calculate OEE and TEEP based on the following data: 5600 hours uptime, 400 hours downtime, expected production based on design of 1000 units per production run, actual production of 850 units per production run, 50 units defective per production run, no demand for 2760 hours per year.
24. Draw a diagram of the OEE and TEEP components based on the previous data.
25. What are the benefits of TPM?

Research Questions

(Always cite your references when answering research questions.)

1. Describe at least one company's transformation to an operator-driven reliability culture.
2. Contrast the positive and negative attributes of a division of labor between operations and maintenance.
3. Research the approaches taken by maintenance management teams to convince maintainers to take a pay cut by reducing overtime when transitioning to an operator-maintainer concept.
4. Discuss the relationship between Just-in-Time/Lean Manufacturing and an operator-driven reliability culture.
5. Give a detailed history of the development of the TPM concept.
6. Compare and contrast TPM and TQM philosophies.
7. Discuss how some companies consider maintenance a non-value-added function.
8. Research how continuous improvement tools — such as Pareto, 5-Why, and FMEA — can be used to reduce operations losses and improve operational effectiveness.
9. Describe the concept of Maintenance Prevention or Design and Early Equipment Management.
10. Research activities performed by different organizations to better organize their workplace.
11. Discuss best practices in labeling and identifying operations and maintenance items as well as assets in creating a visual workplace.
12. Describe how cleaning and proper housekeeping can be an important part of achieving operations and maintenance effectiveness.
13. Research standardization practices that different organizations are using to improve their operations and maintenance programs.
14. Discuss how sustainment and discipline are carried out in an organization in order to effectively manage an operations and maintenance organization.
15. Contrast the positive and negative reasons for measuring OEE in an organization.
16. Contrast the positive and negative reasons for measuring TEEP in an organization.
17. Discuss what facilities are doing to evaluate and improve asset performance.
18. Discuss what organizations are doing to evaluate and improve quality.
19. Research new measures that could be used to evaluate and improve an organization's implementation and sustainment of an operator-driven reliability culture.

Chapter 8
Maintenance Optimization

"Innovative practices combined with true empowerment produce phenomenal results."
Captain Michael Abrashoff,
Former Commanding Officer, USS Benfold,
Author, *It's Your Ship*

Chapter 8: Maintenance Optimization
Chapter Outline

- 8.1 Introduction
- 8.2 Key Terms and Definitions
- 8.3 Understanding Failures and Maintenance Strategies
- 8.4 Maintenance Strategy — RCM
- 8.5 Maintenance Strategy — CBM
- 8.6 Other Maintenance Strategies
- 8.7 Summary
- 8.8 Chapter Assessment
- 8.9 References and Suggested Reading

Chapter 8's Objective … to understand

- What is a failure?
- What is RCM?
- What does it take to implement RCM effectively?
- What CBM technologies are available?
- What are the different maintenance strategies?
- How can you integrate PM and CBM into RCM methodology
- When would RTF be a good maintenance strategy?

MAINTENANCE OPTIMIZATION

Introduction

Maintenance has entered the heart of many organizational activities due to its vital role in:
- Environmental preservation
- Productivity
- Quality
- System reliability
- Regulatory compliance
- Safety
- Profitability

RCM/PM/CBM/RTF

- Central to maintenance is a process called Reliability-Centered Maintenance (RCM)
  - Provides a structured framework for analyzing functions and potential failures
  - Helps in developing an effective maintenance plan by selecting appropriate strategies of PM, CBM, and RTF
- Different Maintenance Strategies:
  - CBM — Attempts to evaluate condition of assets by performing periodic or continuous condition monitoring
  - PM — Planned maintenance of assets based on time or some other interval-type measure
  - RTF — Proactively decide to allow asset to fail without any periodic maintenance

How Do We Best Understand Failure?

The "P-F" Curve

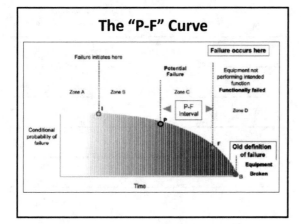

Understanding Failure

The old definition of failure is the end point of this curve (the "B") – the point at which the asset breaks down.

- Failure starts much before this point ("I" – the point at which the asset initiates failure).
- Failure potential increases until reaching Potential Failure ("P" – point at which failure is detected).
- Eventually, Potential Failure turns into a Functional Failure ("F" – point at which asset is unable to perform one or more of its functions).
- Asset continues to operate at a reduced capacity or functionality, until completely breaks down ("B").

"P-F" Interval

The time interval between the "P" and the "F" is known as the "P-F" interval.

- This interval theoretically dictates PM or CBM tasks in order to identify and correct failures in time.
- The challenge is that we don't readily know exactly where "P" and "F" are for every asset or component.
- Therefore, we continually assess condition and operating data to better understand how to set the PM and CBM intervals properly.

MAINTENANCE OPTIMIZATION

What Is RCM?

Reliability-Centered Maintenance

- A systematic and structured process to develop an efficient and effective maintenance plan for an asset that minimizes the probability of failures while ensuring safety and mission compliance.
- Successful implementation of RCM will promote cost-effectiveness, improve asset uptime, and help to better understand the level of risk that the organization is currently managing.
- It has been demonstrated that, in order to eliminate or mitigate the effects of failure modes, the best application of RCM is during design and development phase of the assets.

RCM History

It all started with a Boeing 747:
- In the late 1960s, all airplane owners/operators were required to gain approval of their PM program from the FAA to get certified for operation.
    - The Boeing 747's size and many technological advances led the FAA to take the position that its PM program would be very extensive.
    - So expensive that it may not be possible to operate profitably with this requirement.
- This led the commercial aircraft industry to undertake a complete re-evaluation of their PM strategy.
    - United Airlines led entirely new approach with a decision-tree process for ranking PM tasks necessary to preserve critical aircraft functions during flights.
    - New technique defined/explained in Maintenance Steering Group 1 (MSG1) for 747 and subsequently approved by FAA.

RCM History

Further implemented and instituted in industry:
- RCM principles applied to other aircraft: DC-10, MD-80/90, Boeing 757 / 777, Navy P3/S3, and Air Force F-4J.
- In 1975, DoD directed that the MSG concept be labeled Reliability-Centered Maintenance (RCM) and be applied to all major military systems.
- In 1978, United Airlines produced the initial RCM "bible" under DoD contract.
- RCM development has been an evolutionary process.
  - Over 40 years have passed since its inception, during which RCM has become a mature process.
  - Many M&R leaders have increased RCM awareness and continued its application by optimizing PMs utilizing RCM methodologies.

RCM Principles

- *Principle 1:* The primary objective of RCM is to preserve system function.
- *Principle 2:* Identify failure modes that can defeat the functions.
- *Principle 3:* Prioritize function needs (failures modes).
- *Principle 4:* Select applicable and effective tasks (that achieve one or more of these):
  - Prevent or mitigate failure
  - Detect onset of a failure
  - Discover a hidden failure

RCM Process Elements

The SAE JA1011 standard describes the minimum criteria to which a process must comply to be called **RCM**:
1. What are the functions and associated desired standards of performance of the asset in its present operating context (functions)?
2. In what ways can the asset fail to fulfill its functions (functional failures)?
3. What causes each functional failure (failure modes)?
4. What happens when each failure occurs (failure effects)?
5. In what way does each failure matter (failure consequences)?
6. What should be done to predict or prevent each failure (proactive tasks and task intervals)?
7. What should be done if a suitable proactive task cannot be found (default actions)?

RCM Analysis Process

Although RCM varies a great deal in application, most procedures include the following 9 steps:

1. System selection and information collection
2. **System boundary definition**
3. System description and functional block diagram
4. **System functions and functional failures**
5. **Failure Modes and Effects Analysis (FMEA)**
6. Logic (decision) Tree Analysis (LTA)
7. **Selection of maintenance tasks**
8. Task packaging and implementation
9. **Making the program a living one — continuous improvement**

Bolded steps covered in more detail in subsequent slides

Foundation of Effective RCM

- It all begins with forming the "best" team.
- Selection of RCM team members is key to success.
- The team should include the following:
  - System operator (craft)
  - System maintainer (craft – mechanical / electrical / controls)
  - Operations / Production engineer
  - Systems / Maintenance engineer (mechanical / electrical)
  - CBM/PdM specialist or technician
  - Facilitator
- Actual team members dependent on level of effort required and complexity of specific RCM effort.

Defining System Boundaries

- It will be very easy for a team to try to solve too many problems they are having with their assets.
- Therefore, it is imperative that RCM efforts be properly constrained to minimize this potential and to ensure the effectiveness of its outcome.
- Defining system boundaries helps understand the system as a whole and its functional sub-systems.
  - This step assures that there are no overlaps or gaps between adjacent systems.
  - Define exactly what is approaching or crossing this boundary
    - "Incoming – IN" interfaces
    - "Outgoing – OUT" interfaces
  - This boundary also aids in having a clear record for future reference on the system analyzed with RCM.

System Boundary Example

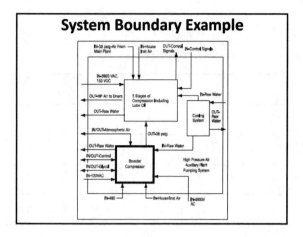

System Functions and Functional Failures

- Because the ultimate goal of RCM is "to preserve system function," it is incumbent upon the RCM team to define and document a complete list of system functions and functional failures.
- The function statement should describe what the system does (functions).
- In theory, we should be able to stand outside of a selected system boundary, with no knowledge of the system components, and define the functions by simply what is leaving the system (OUT interfaces).
- Example function statements:
  - A correct function might be "Maintain a flow of 1000 GPM at header 25," but not "provide a 1000 GPM centrifugal Pump for discharge at header 25."
  - Or "Maintain lube oil temperature ≤ 110°F."

FMEA

- The Failure Modes and Effects Analysis (FMEA) step is the heart of the RCM process.
- FMEA has been used traditionally to improve system design and is now being used effectively for failure analysis that is critical to preserve system function.
- By developing the functional failure – equipment matrix, the connection between function and hardware is established.
- This tool is discussed in detail in Chapter 11.

MAINTENANCE OPTIMIZATION

FMEA Example

FF#	Comp #	Comp. Description	FM#	Failure Mode	FC#	Failure Cause	Failure Effect Local	System	Plant	LTA
1.1	01	6.9 kV Motor	1.01	Bearing seizures	1.1.1	1) Oil contamination 2) Lack of lubrication	Can't supply air to C93A/B	Can't supply air to C93A/B	Loss of high pressure air (HPA)	YES
1.1	01	6.9 kV Motor	1.02	Bearing wear	1.2.1	1) Normal use 2) Oil contamination	Elevated bearing temperature and vibration	Worst case: Can't supply air to C93A/B	Loss of high pressure air (HPA)	YES
1.1	01	6.9 kV Motor	1.03	Windings shorted or open (insulation degradation)	1.3.1	1) Aging and contamination 2) End winding corona	Motor won't run	Can't supply air to C93A/B	Loss of high pressure air (HPA)	YES
1.1	01	6.9 kV Motor	1.04	Loose or cracked surge ring and end turn blocking and ties	1.4.1	Vibration and heat	Elevated winding temperatures that lead to failure	Worst case: Can't supply air to C93A/B	Loss of high pressure air (HPA)	YES
1.1	01	6.9 kV Motor	1.05	RTD fails open	1.5.1	Internal failure	Loss of motor temperature minimum	None	None	NO
1.1	01	6.9 kV Motor	1.06	Air filter clogging	1.6.1	Dirt	Motor runs hot	None	None	NO

Failure Mode Guidelines

Teams can use the following guidelines in accepting, rejecting, or putting aside failure modes for later consideration:

- **Probable Failure Mode:** *Could this failure mode occur at least once in the life of the equipment / plant? If yes, keep it; if no, drop it.*
- **Implausible Failure Mode:** *Does this failure mode defy the natural laws of physics — is it one that just could not ever happen? If no, keep it; if yes, drop it.*
- **Maintainable Failure Mode:** *Can this failure be maintained with a practical approach? If yes, keep it; if no, drop it.*
- **Human Error Causes:** *Is the only way this failure mode can happen is with an unfortunate (but likely) human error? If no, keep it; if yes, drop it.*

Logic (Decision) Tree Analysis

Failure Mode
(1) Evident — Under normal conditions, do the operators know that something has occured?

- Yes → (2) Safety — Does this failure Mode cause a Safety Problem?
- No → (D) Hidden failure — Return to logic tree to ascertain if the failure is an A, B, or C

(2) Safety:
- Yes → (A) Safety Problem — Optional, continue to question 3
- No → (3) Outage — Does this failure Mode result in full or partial outage of the plant?
  - Yes → (B) Outage Problem
  - No → (C) Minor to insignificant economic problem

Maintenance Task Selection

FMEA/RCM efforts are useless without appropriate maintenance task selection for implementation.
- Team knowledge determines the most applicable and cost effective tasks that will eliminate, mitigate, or warn us of failure modes and causes.
- If a maintenance program currently exists, then this task selection is used to improve this program.
- In task selection, the following slide's questions are addressed.

Maintenance Task Selection

- Is age reliability relationship for this failure known? If yes,
  - Are there any applicable Time directed (TD) tasks? If yes, specify those tasks.
- Are there any Condition Directed (CD) tasks? If yes, specify those tasks.
- Are there any Run-to-Failure (RTF) failure modes? If yes,
  - Are there any applicable Failure Finding (FF) Tasks? If yes, specify those tasks.
- Can any of the preceding tasks be ineffective? If no, finalize tasks above as TD, CD, and FF tasks.
- If any tasks may not be effective, then can design modifications eliminate failure mode or effect? If yes, request design modifications.

"Living" RCM Program

- RCM execution is not a one-time event; it's a journey.
- RCM is a paradigm shift in how maintenance is perceived and executed.
- An RCM-based maintenance program needs to be reviewed and updated on a continuous basis.
- A living RCM program consists of:
  - Validating existing program to ensure maintenance decisions made are appropriate.
  - Reviewing current failure history and evaluating maintenance task effectiveness.
  - Making adjustments in maintenance program if needed.
- A living RCM program assures continual improvement and cost-effective operation and maintenance in the organization.
- This highlights the need to establish some effective metrics to know where the program stands.

Benefits of RCM

- The cost and effort to execute RCM can be substantial, dependent on the system.
- The benefits typically outweigh the costs.
  - **Safety:** This is the primary concern of RCM.
  - **Reliability:** RCM's primary goal is to improve asset reliability/availability cost-effectively.
  - **Cost:** Total maintenance costs are reduced due to failures prevented and PM tasks replaced by CBM/PdM.
  - **Documentation:** O&M docs captured and understood.
  - **Equipment/Parts Replacement:** Maximum use from the equipment or system is obtained by extending the life of the facility and its equipment.
  - **Efficiency/Productivity:** Proper type of maintenance is performed only when it is needed.

What Do CBM and PdM Mean?

What Is CBM/PdM?

- Condition-Based (or Predictive) Maintenance (CBM / PdM) is maintenance based on the actual condition (health) of assets obtained from in-place, non-invasive measurements and tests.
- These terms are used both separately and synonymously in industry.
- Any CBM/PdM Program can be characterized by a combination of 3 distinct but related efforts:
  - **Surveillance:** Monitoring machinery condition to detect incipient problems
  - **Diagnosis / Prognosis:** Isolating the cause of the problem and developing a corrective action plan based on its condition and remaining life
  - **Remedy:** Performing corrective action
- Data collection that is both consistent and accurate is essential to all three phases.

152 CHAPTER 8

Data Collection and Analysis

- Asset condition data collected primarily with:
  - Spot readings (route based with portable instruments)
  - Permanently installed data acquisition equipment for continuous online data collection
- Analyzed using multiple methods:
  - Trend Analysis
  - Pattern Recognition
  - Correlation Analysis
  - Tests Against Limits and Ranges
  - Statistical Process Analysis

With Multiple Technologies

There are many different characteristics that can be measured and analyzed to understand asset health, with some more widely used than others:
- **Vibration Analysis**
- **Infrared Thermography**
- **Ultrasonic Testing (contact and airborne)**
- **Lubricant Analysis**
- **Electrical Testing**
- Flow Rates
- Temperature
- Pressure
- Other?

Bolded steps covered in more detail in subsequent slides

Vibration Analysis

- **Vibration** is used to assess the condition of rotating equipment. Additionally, structural problems can be identified through resonance and modal testing.
- Vibration has four fundamental characteristics:
  - **Frequency** is the number of cycles per unit time and is expressed in the number of cycles per minute (CPM) or cycles per second (Hz).
  - **Period** is the time required to complete one cycle of vibration, the reciprocal of frequency.
  - **Amplitude** is the maximum value of vibration at a given location of the machine.
  - **Phase** is the time relationship between vibrations of the same frequency and is measured in degrees.

Vibration Analysis

- Vibration has three key measures:

Measure	Units	Description
Displacement	mils peak-to-peak (p-p)	motion of machine, structure or rotor – relates to stress
Velocity	in / sec	rate of motion, relates to usually component fatigue
Acceleration	in/2 sec or g's	relates to forces present in components

1 mil = 0.001 inch and 1g = 386.1 inches/sec

- Vibration Analysis Categories:
  - Broadband Trending
  - Narrowband Trending
  - Signature Analysis
  - Spectrum Analysis
  - Waveform or Time Domain Analysis
  - Shock Pulse Analysis

Vibration Applications

- With few exceptions, mechanical issues in a rotating machine cause vibration.
- Common problems that produce vibration are:
  - Imbalance of rotating parts
  - Misalignment of couplings and bearings
  - Bent shafts
  - Worn, eccentric, or damaged parts
  - Bad drive belts and chains
  - Damaged / bad bearings
  - Looseness
  - Rubbing
  - Aerodynamic and other forces

Infrared Thermography

- **Infrared Thermography** is used to study everything from individual components of assets to plant systems, roofs, and even entire buildings.
- Infrared inspections can be qualitative or quantitative.
  - Qualitative concerns relative differences, hot and cold spots, and deviations from expected temperatures.
  - Quantitative concerns accurate measurement of the temperature of the target.
  - But don't put too much emphasis on the quantitative side of infrared because temperature-based sensors are better for accurate temperature measurements.

Infrared Thermography

- It is essential that infrared studies be conducted by technicians who are trained in the operation of the equipment and interpretation of the imagery.
- Variables that hurt accuracy and repeatability must be compensated for each time data is acquired.
  - Emissivity (material property based on its ability to display true thermal characteristics vs. those reflected by a nearby item) is an especially key concern that can introduce 5–20% error in measurements.
  - Environmental factors (solar heating and wind effects)
  - Errors can also be introduced due to color of material and material geometry.
- Infrared Thermography is limited to line of sight.

Infrared Applications

IRT can be used very effectively:
- Identify degrading conditions in facilities' electrical systems such as transformers, motor control centers, switchgear, substations, switch yards, or power lines.
- Identify blocked flow conditions in heat exchangers, condensers, transformer cooling radiators, and pipes.
- Verify fluid level in large containers (e.g., fuel storage tanks).
- Identify insulation system degradation in building walls and roofs, as well as refractory in boilers and furnaces.
- Find moisture-induced temperature effects (e.g., roof leaks).
- Determine the thermal efficiency of heat exchangers, boilers, building envelopes, etc.
- Detect buried pipe energy loss and leakage by examining the temperature of the surrounding soil.

Ultrasonic Testing

- **Ultrasonic Testing** is extremely useful in the diagnosis of mechanical and electrical problems as well as gas, liquid, or vacuum leaks.
- Ultrasonic testing instruments are typically portable handheld devices whose electronic circuitry converts a narrow band of ultrasound (between 20 and 100 kHz) into the audible range so that a user can recognize the qualitative sounds of operating equipment through headphones.
- Ultrasonic monitoring requires minimal training, and the instruments are inexpensive.

Ultrasonic Applications

Some of the most common applications of ultrasound detection are:
- **Leak and flow detection in pressure and vacuum systems** (e.g., boiler, heat exchanger, condensers, chillers, vacuum furnaces, specialty gas systems, steam traps, integrity of seals and gaskets in tanks/piping systems)
- **Mechanical equipment inspections** (bearings, pump cavitation, valve analysis)
- **Electrical equipment inspections** (arcing, tracking, corona)

Oil Analysis

The objective of oil (lubricant and wear particle) analysis is to determine:
- An asset's mechanical wear condition
- Lubricant condition
  - Bad lubricating oil is either discarded or reconditioned through filtering or by replacing additives.
  - Oil is changed (or reconditioned) based on
    - Operating time basis
    - Condition based via oil analysis
- If the lubricant has become contaminated

Standard Analytical Oil Tests

- Visual and Odor
- Viscosity
- Water (moisture)
- Wear Particle Count
- Total Acid Number (TAN)
- Total Base Number (TBN)
- Spectrometric Metal Analysis
- Infrared Spectroscopy
- Analytical Ferrography
- Foaming
- Rust Prevention
- Rotating Bomb Oxidation

Oil Sampling & Frequency

- Oil samples must be collected safely and in a manner that will not introduce dirt and other contaminates into the machine or system, or into the sample.
- It may be necessary to install permanent sample valves in some lubricating systems.
- The oil sample should be representative of the oil being circulated in the machine. Therefore:
  - Collect samples from the same point in the system to ensure consistency in the test analysis.
  - Collect from a mid-point in reservoirs and upstream of the filter in circulating systems.
  - Use clean sample collection bottles and tubing to collect the sample.

Oil Contamination Program

- A basic oil contamination control program can be implemented in three steps:
  1. Establish the target fluid cleanliness levels for each machine fluid system.
  2. Select and install filtration equipment (or upgrade current filter rating) and contaminant exclusion techniques to achieve target cleanliness levels.
  3. Monitor fluid cleanliness at regular intervals to achieve target cleanliness levels.
- It is a good practice to establish a quality control program for incoming oil; therefore, set up a minimum oil cleanliness standard (ISO 4406) for all oils, new or old, before they are used in the machines.

Oil Analysis Applications

Both mechanical and electrical issues can be detected with oil analysis:
- Bearing or other rotating component wear
- Water or moisture intrusion
- Oil breakdown
- Electrical discharge (as with transformer oil analysis)

Electrical Testing

Electrical Testing (condition monitoring) encompasses several technologies and techniques used for a comprehensive system evaluation.
- Electrical equipment represents a major portion of a facility's capital investment.
- From the power distribution system to electric motors, efficient operation of the electrical systems is crucial to maintaining operational capability of a facility.
- Monitoring key electrical parameters provides ability to detect and correct electrical faults (high resistance connections, phase imbalance, and insulation breakdown).
- Because faults in electrical systems are seldom visible, these faults are costly due to increased electrical usage and increased safety concerns.

Electrical Testing Categories

Electrical equipment evaluation can be divided into two categories:
- **Online Monitoring/Testing** measures any aspect of electrical equipment while it is in service and operating.
- **Offline Testing** is performed by inducing voltage or current into equipment and taking electrical measurements, which requires that the equipment be de-energized and completely isolated from its normal circuit.
- **Both** are very valuable in evaluating an electrical system. In most cases, they detect different types of faults or potential problems, meaning much of online monitoring/testing cannot replace offline testing and vice versa.

Types of Electrical Testing

Several of these types also used for acceptance testing and certification for new systems:
- Motor Current Readings
- Motor Current Signature Analysis (MCSA)
- AC High Potential Testing (HiPot)
- Surge Comparison Testing
- Conductor Complex Impedance
- Megohmmeter Testing
- Polarization index
- Time Domain Reflectometry
- Radio Frequency (RF) Monitoring
- Power Factor and Harmonic Distortion

Electrical Testing Applications

Specific electrical assets that can be monitored by CBM technologies are:
- Electrical Distribution Cabling
- Switchgear and Controllers
- Transformers
- Motors
- Generators

Other Miscellaneous NDT

- Non-Destructive Testing (NDT) evaluates material properties and quality of expensive components or assemblies without damaging the product or its function.
- There are various NDT techniques:
  – Radiography
  – Ultrasonic Testing (Imaging)
  – Magnetic Particle Testing (MPT)
  – Hydrostatic Testing
  – Eddy Current Testing

Why CBM?

- Warns about most problems in time to minimize:
  - Unexpected failure
  - Risk and consequences of collateral damage
  - Adverse impact on safety, operations, & environment
- Increase equipment utilization, life, uptime, availability, and worker safety:
  - Awareness of asset condition
  - Likelihood that components operate to optimum lifetime
  - Reduce requirements for emergency spares

Why CBM?

- Reduce O&M cost — both parts and labor.
  - Reduce significant amount of calendar/run-based PM.
  - Minimize cost and hazard to an asset that result from unnecessary overhauls, dis-assemblies, and PM inspections.
  - Increase energy savings.
- Provide vital information for continuous improvement, work, and logistic planning.

What CBM Cannot Do

- CBM is not a "silver bullet." Some potential failures, such as fatigue or uniform wear on a blower fan are not easily detected with condition measurements.
- In other cases, sensors may not be able to survive in the environment; measurements to assess condition may be overly difficult and may require major asset modifications.
- However, CBM cannot:
  - Eliminate defects and problems, or stop assets from deteriorating.
  - Eliminate all preventive maintenance.
  - Reliably and effectively warn of fatigue failures
  - Reduce personnel or produce a major decrease in lifetime maintenance costs without a commitment to eliminating defects and chronic problems.

What about PM?

Preventive Maintenance (PM)

- Preventive maintenance refers to a series of actions that are performed on an asset on schedule.
  - That schedule may be either based on time or based upon machine runtime or the number of machine cycles.
  - These actions are designed to detect, preclude, or mitigate degradation of a system and its components.
- The goal of a preventive maintenance approach is to minimize system and component degradation and thus sustain or extend the useful life of the asset.

The Objective of PM

The objective of preventive maintenance (PM) can be summarized as follows:
- Maintain assets and facilities in satisfactory operating condition
- Perform maintenance to prevent failure from occurring
- Record asset health condition for analysis

PM Myths & Practices

- Most existing PM programs cannot be traced to their origins. For those that can, most are unlikely to make sense.
- The following reasons are usually the ones given for a PM program:
  - OEM recommendations
  - Experience
  - Failure prevention
  - Brute force
  - Regulations
- Rather than using the preceding reasons, a good PM program should be based on a FMEA / RCM analysis.

10% Rule of PM

- A PM plan must be executed per schedule.
- The best practice is to use the 10% rule of PM.
- This rule should apply to all PMs, but at a minimum with critical assets.
- Organizations that have implemented the 10% rule have been found to have increased reliability of the assets due to a consistent and disciplined approach.

If we are performing Preventive Maintenance on an asset that continues to fail, we are in a reactive maintenance mode.
The PM plan should be reviewed and adjusted.

What Is an RTF strategy?

Run-to-Failure (RTF)

RTF is a maintenance strategy where a conscious decision is made to allow specific assets/systems to fail without any PM or CBM against them.

- This strategy is not the same as reactive maintenance.
- In reactive maintenance, an organization does not have a structured maintenance program, which would include elements of PM, CBM, and RTF spread throughout the facility, with each asset/system having its own specific maintenance strategy.
- For assets where cost and impact of failure is less than cost of preventive (PM and CBM) actions, RTF may be an appropriate maintenance strategy because it is a deliberate decision based on economical effectiveness.

Run-to-Failure (RTF)

- Additionally, organizations must have a plan to repair the failure, if and when it happens.
- Many times, we do consider and accept RTF for specific non-critical assets or components.
  - It is imperative to document that RTF was selected proactively.
  - Documentation minimizes excitement when RTF failure occurs.
  - A deliberate economical decision not to have a PM program for that asset.

Which All Leads to a Strategy

- All assets, or at a minimum critical assets, should have an asset management strategy established and documented.
- An asset management strategy should have:
  - Maintenance strategies selected and why (basis)
    - List by subassembly or components
    - Asset hierarchy structure identified
  - Suggested spare parts needed (when and how many)
  - List of PM and CBM actions/tasks/routes
  - Major repair plans
  - Operating/troubleshooting guidelines or procedures
  - Any specific qualification/certification needs for maintenance personnel

Maintenance Optimization Measures of Performance

- RCM implementation % (overall equipment)
- Involvement of ops and maintenance personnel in RCM efforts
- % CBM (overall maintenance)
- % PM (overall maintenance)
- % RTF (overall equipment)

Summary

- Reliability-Centered Maintenance (RCM) is a maintenance improvement approach focused on:
  - Identifying and establishing the operational, maintenance, and design improvement strategies
  - Managing risks of asset failure most effectively
- The RCM process determines what must be done to ensure that assets continue to do what their users need them for in a certain operating context.
- RCM analysis provides a structured framework for analyzing functions and potential failures of assets.
- RCM is a maintenance /PM plan optimizing strategy. However, RCM analysis provides the maximum benefits during the asset's entire life.

Summary

- Condition-Based Maintenance (CBM) determines what must be done to ensure assets continue to function cost-effectively based on actual operating environment.
- CBM is based on using real-time data to assess the condition of the assets utilizing predictive maintenance technologies.
  - The data and its analysis help us to make better decisions to optimize maintenance resources.
  - CBM will determine the equipment's health, and act only when maintenance is actually necessary.
- CBM attempts to predict impending failure based on actual operating data instead of traditional PM, generally eliminating unnecessary maintenance performed.

Summary

- Preventive Maintenance (PM) is the basic asset (maintenance) strategy that many organizations use to begin their formal maintenance program.
  - Most PM programs are either calendar time-based or operations runtime-based.
- Run-to-failure (RTF) is another economically valid strategy for specifically identified assets/systems.
  - This strategy must be deliberately selected for non-critical assets only.
  - It should be documented and planned with the right level of support, such as spare parts.

Chapter 8 Self-Assessment Questions

1. What is RCM? How did it get its start? Tell a little about RCM's history.
2. Which Standards Development Organization (SDO) developed RCM Standard JA1011?
3. Describe the 4 principles of RCM. What is the key objective of RCM analysis?
4. During what phases of asset development do we get the maximum benefit of a RCM analysis? Why?
5. Describe the 9-step RCM analysis process.
6. Which type of failure mode is not evident to the asset operator?
7. What are the benefits of RCM ?
8. What is meant by CBM and PdM? What methods are used to perform these?
9. What is the difference between diagnostic and prognostic analysis?
10. What is velocity analysis? With which CBM technology is it associated? What does a peak at twice rotational speed indicate?

MAINTENANCE OPTIMIZATION

Chapter 8   Comprehension Assessment Questions

1. Where does maintenance provide a vital role within an organization and its activities?
2. What is meant by the PF interval?
3. Draw the PF interval diagram.
4. What are 3 reasons that one might choose to perform PM or CBM?
5. What are the essential questions the RCM process must answer?
6. What are the tangible actionable options that result from an RCM analysis?
7. Which team members are key to success in an RCM program?
8. What are the benefits of using a facilitator within an RCM effort?
9. What is an FMEA?
10. What are the different types of failure modes that should be excluded from further analysis within the FME a process?
11. At what point during the RCM analysis process should the RCM team review the current PM or maintenance task structure in detail?
12. What are the key elements of a living RCM program?
13. What are the potential O&M savings of a balanced RCM program?
14. What is the major recent advance in technology that has made CBM/PdM a reality?
15. What are the three phases of CBM/PdM programs?
16. What are the two basic methods used to collect CBM/PdM data?
17. What are the different CBM/PdM analysis methods used to determine the condition of the asset?
18. What are the four fundamental characteristics of vibration?
19. What are the key measures used to evaluate the magnitude of vibration?
20. What types of failures or problems will manifest themselves as vibration?
21. What are the three major types of vibration analyses?
22. What is the difference between a qualitative and quantitative infrared inspection and why is this difference important?
23. Why is it essential that infrared studies be conducted by technicians who are trained in the operation of the equipment and interpretation of the imagery?
24. Give at least 4 examples of equipment characteristics/issues detected with Infrared Thermography.
25. What is emissivity and why is it important to Infrared Thermography?
26. What is the typical range of Ultrasonic testing in kHz?
27. What is corona and what does it have to do with Ultrasonic testing?
28. List at least 4 applications of Ultrasonic testing.
29. What are the 3 objectives of Oil Analysis?
30. List at least 5 of the possible Oil Analysis or Oil sampling tests that may be performed.
31. What are the 3 basic steps to implement a basic oil contamination control program?
32. Describe the 2 basic categories of Electrical Equipment Evaluation.
33. List 5 of the different Electrical Tests that can be used?
34. List 5 of the benefits of having a CBM/PdM program?
35. Why is CBM not a "Silver Bullet"?

36. What is a PM?
37. What are the objectives of a PM program?
38. What is the PM program typically based upon?
39. What should a PM program be based upon?
40. What are the risks associated with poor workmanship when performing PM tasks?
41. What is meant by the "10% Rule of PM"?
42. What is RTF?
43. What is the difference between a RTF strategy and Reactive Maintenance?

Research Questions

(Always cite your references when answering research questions.)

1. Discuss a detailed history of the development and implementation of RCM.
2. Describe the theory behind what is known as the "P-F Curve."
3. Compare the differing viewpoints on whether one should wait as close to the "F" part of the "P-F Curve" versus correcting the problem as soon as it is identified (which is closer to the "P" part of the "P-F Curve."
4. Discuss details about the military standards related to RCM and its application on major military systems.
5. Describe what is meant by the "age exploration" process.
6. Evaluate at least one specific company's implementation of RCM within the design process.
7. Discuss the return on investment that companies have gotten from implementation of RCM.
8. Discuss the relationship between human error and RCM.
9. Summarize the theory behind logic or decision tree analysis.
10. Describe in detail any of the CBM/PdM technologies not discussed in detail within this book.
11. Discuss any new CBM/PdM technologies not listed in this book.
12. Compare the advantages and disadvantages of "smart" CBM/PdM data analysis software.
13. Discuss the disadvantages of using only a CBM program as your maintenance strategy.
14. Compare the advantages and disadvantages of using OEM recommendations as a basis for your O&M strategy.
15. Describe at least five regulations related to maintenance that could be improved based on CBM.
16. Compare the advantages and disadvantages of using an online continuous monitoring system versus route-based CBM/PdM data collection.
17. Compare an alternative PM compliance rule or metric to the "10% Rule of PM."
18. Evaluate a specific company's implementation of an RTF strategy as a major part of their overall maintenance program.

Chapter 9
Managing Performance

*"You cannot manage
something you cannot control,
and you cannot control
something you cannot measure."*
—Peter Drucker

Chapter 9: Managing Performance

9.1 Introduction
9.2 Key Terms and Definitions
9.3 Identifying Performance Measures
9.4 Data Collection and Data Quality
9.5 Benchmarking and Benchmarks
9.6 Summary
9.7 Chapter Assessment
9.8 References and Suggested Reading

Chapter 9 Objectives

- What to measure and why to measure performance
- Differences between lagging and leading indicators
- Key performance indicators
- A balanced scorecard
- The challenges of data collection
- The importance of data quality and integrity
- Benchmarks and benchmarking

Introduction

- An organization must measure and analyze its performance.
  - To make the improvements needed for staying in business
  - In a competitive market place
- Performance measures must be derived and aligned with organization's goals and strategies.
  - Centered on critical information and data related to the key business processes and outputs
  - Focused on improving results

Performance Indicators Are Key

- Performance indicators, also known as metrics, are measurable characteristics of products, services, and processes related to the business.
  - Used to track and improve performance
  - Comprehensive set tied to business activities and customers based on long-term and short-term goals
  - Need to be constantly reviewed and aligned with new or updated goals as part of its strategic plan to make lasting improvements to key drivers
- Processes are designed to collect information and reduce for easy dissemination and analysis.

Data, Data, & More Data

- Data is needed for process improvement and performance measurement for:
  - Products and services
  - Asset performance
  - Operations and maintenance
- Data can be easy or difficult to collect, which means organizational emphasis should be placed on data quality.

Data, Data, and More Data

- Data is analyzed to determine:
    - Trends
    - Causes and effects
    - Underlying reasons for certain results that may not be evident without analysis
- Data also used to serve a variety of purposes:
    - Planning and projections
    - Performance reviews and assessments
    - Operations and maintenance improvements
    - Comparing an organization's performance with the "best practices" benchmarks

The Value of Metrics

The value of metrics is in their ability to provide a factual basis in the following areas:
- Strategic feedback to show present status of the organization from various perspectives
- Diagnostic feedback of various processes to guide improvements on a continuous basis
- Trends in performance over time as metrics are tracked
- Feedback around measurement methods themselves in order to track the correct metrics

Performance Measurement Benefits

Benefits related to performance measurement:
- Accountability: Documenting progress toward achievement of goals and objectives
- Resource/Budget Justification: Long-term planning to justify resource/budget allocation
- Ownership and Teamwork: Provides for more employee participation in problem solving, goal setting, and process improvement activities
- A Common Language: Giving employees a common language to communicate and share knowledge of management strategies and decisions

What Does All This Have to Do with Maintenance and Reliability?

In most businesses, success is easily measured by looking at the bottom line – the profit.
- What's the bottom line for M&R for business?
- In simple terms, businesses generate profit by selling goods and services and minimizing costs.
- Customers generally demand value such as timeliness, quality, price, and return on investment (ROI).
- M&R metrics should reflect value in terms of timely maintenance (availability of assets), quality of service (minimum rework), and controlling costs.

How Do We Identify the "Right" Performance Measures?

Measuring the "Right" Things

- It is often said that "what gets measured, gets done."
- Getting things done, through people, is what management is all about.
- Measuring things that get accomplished and the results of their effort is an essential part of successful management.
- **Caution!** Too much focus on measurements or the wrong kinds of measurements may not be in the best interest of the organization.

Metrics Development Process

- First, involve those responsible for work being measured because they are most knowledgeable.
- Once these people are identified and involved:
  1. Identify critical work processes and customer requirements.
  2. Identify critical results desired and align them to customer requirements.
  3. Develop measurements for the critical work processes or critical results.
  4. Establish performance goals, standards, or benchmarks.

KPIs

- The vital few measurements important for measuring business performance are called Key Performance Indicators (KPIs)
- Three criteria should be considered:
  - The performance measures should encourage the right behavior.
  - They should be difficult to manipulate to "look good."
  - They should not require a lot of effort to measure.
- An example of M&R KPIs is shown in textbook.

Apply the SMART Test

The SMART test can be used to ensure the quality of a particular performance metric:
- **S**pecific: Should be clear, focused, and easily interpreted to avoid misinterpretation. Include assumptions and definitions.
- **M**easurable: Can be quantified and compared to other data, allowing meaningful statistical analysis. Avoid "yes/no" measures except in limited cases.
- **A**ttainable: Achievable, reasonable, and credible under expected conditions.
- **R**ealistic: Fits into organization's constraints and is cost-effective.
- **T**imely: "Do-able" data available within needed time.

Leading vs. Lagging Indicators

- Leading indicators are forward looking so they help to manage the performance of an asset, system, or process.
- Lagging indicators tell how well we did in managing that performance.

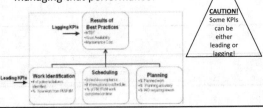

CAUTION! Some KPIs can be either leading or lagging!

The "Balanced Scorecard"

The Balanced Scorecard is a strategic management approach developed in the early 1990s by Dr. Robert Kaplan of Harvard Business School, and Dr. David Norton.

"The balanced scorecard retains traditional financial measures. But financial measures tell the story of past events, an adequate story for industrial age companies for which investments in long-term capabilities and customer relationships were not critical for success. These financial measures are inadequate, however, for guiding and evaluating the journey that organizations must make to create future value through investment in customers, suppliers, employees, processes, technology, and innovation."

"Balanced Scorecard" Perspectives

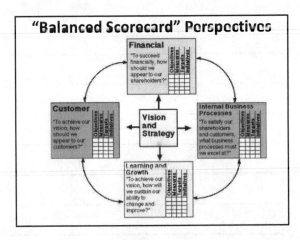

Why Should We Worry about Our Data Collection and Our Data Quality?

Performance Measurement and Data

- Data is the key ingredient in measuring performance.
- Another key challenge is data collection and availability of quality data on a timely basis.
- Major factors in establishing are:
  - Cost of data collection
  - Data quality, completeness, and timeliness
  - Extrapolation from partial coverage
  - Matching measures to their purpose
  - Understanding extraneous influences in the data
  - Use of measures in allocation of funding
  - Responsibility for measures and limited control over the process
  - Benchmarking and targets

Data Collection System

An efficient and effective data collection system is needed to ensure availability of quality data; therefore, a data collection system should:
- Identify data to be collected and how much (including the population from which data will come and the length of time over which to collect data).
- Identify charts and graphs to be used, frequency of charting, various types of comparisons to be made, and methodology for data calculation.
- Identify characteristics of data to be collected.
- Identify if existing data sources can be utilized or if new data sources need to be created for new or updated measures of performance (all data sources need to be credible and cost effective).

What Is the Difference Between Benchmarking and Benchmarks?

What Is Benchmarking?

Benchmarking is the process of identifying, sharing, and using knowledge and best practices.
- It focuses on exploiting top-notch approaches rather than merely measuring best performance.
- Finding, studying, and implementing (internal or external) best practices provide the greatest opportunity for gaining a strategic, operational, and financial advantage.
- Informally, benchmarking could be defined as the practice of being modest enough to admit that others are better at something, and wise enough to learn how to match and even surpass them.

What Is Benchmarking?

Benchmarking is commonly misperceived as simply number crunching, site briefings and industrial tourism, copying, or spying.
- It shouldn't be taken as a quick and easy process.
- It should be considered an ongoing process as a part of continuous improvement.
- It readily integrates with strategic initiatives such as continuous improvement, re-engineering, and total quality management.
- It is also a discrete process that delivers value to the organization on its own.

The Benchmarking Process

1. Conduct internal analysis.
2. Compare data with available benchmarks.
3. Identify gaps in a specific area.
4. Set objectives and define scope.
5. Identify benchmarking partners.
6. Gather information.
   a. Research and develop questionnaire.
   b. Plan benchmarking visits.
7. Distill the learning (compile results).
8. Select practice to implement.
9. Develop plan and implement improvements.
10. Review progress and make changes if necessary.

What Is a Benchmark?

- A benchmark refers to a measure of best practice performance whereas benchmarking refers to the actual search for the best practices.
- A benchmark is a standard, or a set of standards, used as a point of reference for evaluating performance or quality level.
- Benchmarks may be drawn from an organization's own experience, from the experience of others in the industry, or from regulatory requirements.

Summary

Performance measurement is a means of assessing progress against stated goals and objectives in a quantifiable and unbiased way.

- It brings with it an emphasis on objectivity, consistency, fairness, and responsiveness.
- It functions as a reliable indicator of an organization's health and can have an immediate and far-reaching impact.
- Its impact on an organization can be both immediate and far-reaching.
- It asks "What does success really mean?"

Summary

Performance indicators are leading or lagging.
- Leading indicators measure the process and predict changes and future trends.
- Lagging indicators measure results and confirm long-term trends.
- Whether an indicator is a leading or lagging indicator depends on where in the process the indicator is applied.
- Lagging indicators of one process component can be a leading indicator of another process component.

Summary

- A benchmark is a measure of best practice performance.
- Benchmarking refers to the search for the best practices. The process should:
  - Yield the benchmark performance
  - Emphasize how can we implement the best practice
  - Achieve superior results

Summary

Successful performance measurement systems:
- Comprise a balanced set of a limited vital few measures.
- Produce timely and useful reports at a reasonable cost based on quality accurate data.
- Disseminate and display information easily shared, understood, and used by all in the organization.
- Help to manage and improve processes and document achievements.
- Support an organization's core values and its relationship with customers, suppliers, and stakeholders.

Chapter 9 Self-Assessment Questions

1. Why do we need a performance measurement system? What are the benefits of such a system?
2. What are the benefits of benchmarking?
3. Explain what is meant by a "World Class" benchmark.
4. What are the key attributes of a metric?
5. Explain leading and lagging metrics.
6. What types of metrics show results?
7. Explain the Balanced Scorecard model.
8. Explain the different types of benchmarking. What are the benefits of external benchmarking?
9. Discuss data collection and quality issues. How can we improve data quality?
10. List five metrics that can be used to measure overall plant level performance of maintenance activities. Discuss the reason for your selection.

Chapter 9 Comprehension Assessment Questions

1. With what must performance measures be derived and aligned within an organization?
2. What should a comprehensive set of metrics be based upon?
3. What are the key components of value that customers generally demand and how do they relate to maintenance reliability?
4. What are the major benefits of performance measurement?
5. What is a benchmark?
6. What is a metric?
7. What are KPIs?
8. What is the first step in developing metrics?
9. What are the basic steps needed in order to develop metrics?
10. What is meant by the acronym SMART and give specifics about each of its individual acronym letters?
11. What are some examples of maintenance & reliability related "leading" metrics?
12. What are some examples of maintenance & reliability related "lagging" metrics?
13. Give an example of a metric that could be either a "lagging" or "leading" metric and explain how.
14. Choose one of the categorical perspectives of the balanced scorecard model and list associated maintenance & reliability related metrics.
15. Give at least five of the major factors that should be used in establishing a performance measurement system.
16. What are the major things that a data collection system should do?
17. What is benchmarking?
18. List the steps recommended for successfully implementing a benchmarking initiative.
19. What are the major categories of code of conduct that should be considered when benchmarking?
20. What are the key characteristics of a successful performance management system?

Research Questions

(Always cite your references when answering research questions.)

1. Compare differing viewpoints on using few verses many metrics within an organization.
2. Describe a specific company's metric development process.
3. Discuss an alternative to the SMART test for evaluating the quality of metrics.
4. Explain specific views regarding "leading" and "lagging" metrics.
5. Compare the advantages and disadvantages to using the Balanced Scorecard model.
6. Discuss challenges that companies sometimes have in creating an efficient and effective data collection system.
7. Explain what some companies are doing to improve the data quality piece of their data collection system.
8. In this world of multi-site and multinational organizations, describe the positive elements of internal benchmarking that allows the use of external benchmarking components.
9. Evaluate any specific legal challenges (in court) regarding benchmarking between competing organizations.
10. Describe the approach used by the Society of Maintenance and Reliability Professionals in creating standardized metrics in maintenance and reliability.

Chapter 10: Workforce Management

*"What I hear, I forget. What I see, I remember,
What I do, I understand."*
- Kung Fu Tzu (Confucious)

Chapter 10: Workforce Management

10.1 Introduction
10.2 Key Terms and Definitions
10.3 Employee Life Cycle
10.4 Understanding the Generation Gap
10.5 Communication Skills
10.6 People Development
10.7 Resource Management and Organization Structure
10.8 Measures of Performance
10.9 Summary
10.10 Chapter Assessment
10.11 References and Suggested Reading

Chapter 10's Objective ... to understand:

- Deming's view of improving organizational effectiveness
- Employee life cycles
- Generation gaps and the aging workforce
- Communication issues
- People development-related issues
- Training types and benchmarks
- Diversity challenges
- Challenges in managing the workforce, including organization structures and outsourcing

Introduction

- People make "it" (the business purpose) happen – they get things done.
- Without the people available with the right skills, great plans and the best processes can't be implemented effectively.
- Developing the workforce and empowering them to give their best key to the difference between an ordinary company and a great organization.

Introduction (cont'd)

- The right processes and technology must be in place to nurture and harness the potential of human capital.
- This topic is an important challenge in today's global economy and demographically diverse workforce.
- Maintenance and reliability processes are no different than any other processes in the workplace.

Introduction (cont'd)

Dr. W. Edwards Deming, quality guru, listed 14 principles for improving effectiveness.

- Most relate to the workforce, including management and their role in acquiring, preparing, and educating the workforce as well as improving the processes to get productivity gains.
- These principles can be applied to any organization, small to large, and to any industry (including M&R).

CHAPTER 10

Deming's 14 Principles

1. Create constancy of purpose towards improvement.
2. Adopt the new philosophy.
3. Cease dependence on inspection.
4. End the practice of awarding business on the basis of price tag (minimum cost).
5. Improve constantly and forever.
6. Institute training on the job.
7. Institute leadership.

Deming's 14 Principles (cont'd)

8. Drive out fear.
9. Break down barriers between departments.
10. Eliminate slogans and exhortations.
11. Eliminate arbitrary numerical targets.
12. Permit pride of workmanship.
13. Institute education and self-improvement.
14. The transformation is everyone's job.

What Is an Employee Life Cycle?

WORKFORCE MANAGEMENT 187

Employee Life Cycle

- Employees are one of an organization's largest expenses.
  - Unlike other major capital costs such as buildings, machinery, and technology
  - Human capital is highly volatile
- Employee life cycles cover steps employees go through from hiring until leaving:
  - Hire
  - Inspire
  - Admire
  - Retire

The Role of Management

Managers are placed into key positions to reduce volatility by decreasing the overall life cycle cost of employees in the organization.
- Motivate newly-hired employees to stay and add value to the organization.
- Keep employees motivated, happy, and productive, thereby reducing employee turnover.
- Ensure that employees understand the key purpose of the organization (e.g., Disney example – customer vs. guest).

Why Should I Worry about Generation Gaps?

Understanding the Generation Gap

- Shortage of skilled industrial workers poses a serious threat to industrial competitiveness.
  - Those who operate and maintain assets/systems
  - Demographics: Most veterans have left the workplace and Baby Boomers are close to retirement.
- William Strauss & Neil Howe's *Generations*:
  - Silent (born 1925-1942)
  - Baby Boomers (born 1943-1960)
  - Generation X (born 1961-1981)
  - Generation Y (born 1982-2001)

Generation Gap Differences

Differences between generations can create many challenges in the workplace.

- These can be both negative and positive, relating to variations in perspective and goals as a result of generational differences.
- Organizations can't assume these different generations will understand each other.
- Organizations need to understand and value these generational differences and perspectives, turning those negatives into positives.

Closing the Generation Gap

All co-existing generations in the workplace need to understand and value each other, even when their perspectives and goals are different.

- As Silents, Baby Boomers, Gen Xers, and Gen Yers intersect in the workplace – their attitudes, ethics, values, and behaviors inevitably collide.
- Organizations need to leverage multi-generational perspectives to their benefit.
- Organizations must master tools and strategies for communicating across generations to tap the best that each generation and individual brings.

Why Is Communication Important?

The Importance of Communication

- People in organizations typically spend >75% of their time in interpersonal work situations.
    - No surprise that poor communication is at the root of a large number of organizational problems.
    - Effective communication all about conveying messages to others clearly and unambiguously.
    - It's also about receiving information that others send to us, with as little distortion as possible.
- Effective communication exists between two people when receiver interprets and understands sender's message in same way sender intended.

The Communication Process

Problems can occur at every stage of any organization's communication process:

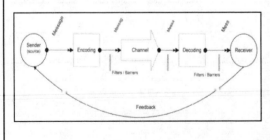

The Importance of Listening

- Listening is one of the most important skills.
  - How well we listen has major impact on our job effectiveness and quality of our relationships.
- We listen to:
  - Obtain information
  - Understand
  - Enjoy
  - Learn
- There are three basic listening modes:
  - Competitive listening
  - Passive or attentive listening
  - Active listening

Become an Active Listener

- Five components of Active Listening:
  - Pay attention
  - Show that we are listening
  - Provide feedback
  - Defer judgment
  - Avoid negative mannerisms but respond
- By becoming better listeners, we will improve our productivity, as well as our ability to:
  - Influence
  - Persuade
  - Negotiate
  - Avoid conflict and misunderstandings

The Value of Meetings

- Meetings are one of the most common forms of communication in the office.
- Meetings take up an ever-increasing amount of employees' and, particularly, managers' time.
- Meetings are wonderful tools for generating ideas, expanding on thoughts, and managing group activity.
  - How often have you thought: "What a waste of time, I could be doing something better!"
  - To ensure meetings are successful, the meeting leader should establish meeting rules.

Effective Meetings

To ensure a meeting is successful:
- Issue an agenda.
- Start the discussion and encourage active participation.
- Work to keep the meeting moving at a comfortable pace (not moving too fast or too slow).
- Summarize discussion and recommendations at the end of each logical break in meeting.
- Ensure all participants receive minutes promptly.

Why Do People (Workforce) Need to Be Developed?

People Development

- Developing people should be viewed as investments.
- If the cost of replacing an employee is known, then it's easy to conclude that getting the most from employees just makes good business sense.
  - Are they performing to the best of their abilities?
  - What training would help them perform better?
  - Can job rotations or on-the-job training (OJT) enhance employees' skills?
  - Could tasks be automated, allowing people to grow in other areas?
  - Are employees aware of how their work fits into overall organizational goals?

Job Task Analysis (JTA)

- Before employees can be trained, we must identify what they need to learn.
- JTA is the foundation of training program success.
    - Purpose is to establish and document skills required for performing jobs effectively, which can help with selection of employee, compensation, performance appraisal, and training.
    - Determines in detail the duties and requirements for a given job with relative importance of each.
    - A job is a collection of tasks and responsibilities that are assigned to an employee.

JTA Methods and Elements

- Several methods may be used individually or in combination to perform JTA:
    - Review of job classification systems, task inventories, and checklists
    - Incumbent and supervisor interviews and logs
    - Observations, expert panels, and structured questionnaires
- Multiple elements of this JTA
    - Duties and tasks
    - Environment, tools, and equipment
    - Relationships and requirements

Skills Development

- Numerous studies have shown that over 50 % of equipment failures result from human error.
- Maintenance personnel skill levels in most organizations are well below what industry would classify as acceptable.
- Issues facing industry include:
    - Literacy level (even basic math and reading skills)
    - General trend of lack of interest in O&M-related work
    - Many organizations have eliminated apprenticeship programs
    - Today's assets and systems are increasingly complex

Training Program Development

- A well-developed training program can provide a solution to this skills shortage.
- Training curriculum should include:
  - Regulatory and safety requirements
  - Technical requirements (asset/system-specific, repair techniques and technology, professional development)
  - Organization-specific (process-related)
  - Certifications and qualifications (industry-standard)
- Training benchmark:

Training Benchmark	Low	High	Best of the Best
Percentage of Overall Budget — Payroll	0.5	6.4	4.5
Expenditure per Employee in $	650	5,000	4200
Training Hours / Employee	5	110	96

How We Manage Resources with an Organization Structure Doesn't Really Matter, Does It?

Resource Management

- Developing people and managing them to be productive is key to a successful business.
- Successful workforce management hinges on:
  - Aligning the workforce with the business strategy
  - Attracting, developing, and retaining key talent
  - Managing diversity
  - Designing the best organization structure for integration of M&R functions
  - Succession planning
  - Developing a leadership culture
  - Establishing and maintaining learning environment
  - Creating a flexible work environment

194   CHAPTER 10

Different M&R Org Structures

There are 3 organization structure types which could be used to set up an M&R organization:
- Centralized
- Decentralized
- Hybrid

Centralized		Decentralized	
Advantages	Disadvantages	Advantages	Disadvantages
Standardized practices	Less responsive to individual units / department	Strong ownership	Difficult to build specialized skills
Enterprise focus — objectives aligned with organization	Lack of ownership	Very responsive to individual area	Difficult to prioritize by facility/department
Efficient use of resources			Sub-optimum use of tools
Easy to build specialized skills			

Different M&R Org Structures

Typical M&R Organization Structure

```
          Capacity Assurance
            (Maintenance)
                 |
   ┌─────────┬───┴────┬──────────┐
Maintenance  M&R    P/S +      Resource
           Engineering Material   Mgmt
```

M&R Functions and Structure

In designing the organization structure, it is important to address the following key M&R functions:
- PM program development
- Execution of PM and CM tasks
- Planning and scheduling
- Material—spares including tools availability
- CBM/PdM and specialized skills, e.g., laser alignment
- Failure elimination and reliability planning
- Designing for reliability
- Resource management and budgeting
- Workforce development

Outsourcing Maintenance

- Outsourcing maintenance activities good practice, particularly with large variations in workload.
- Scheduled shutdowns and major outages may make it economical to contract out work.
- Augmenting maintenance staff with interns and co-op's another cost-effective good practice.
- Some of the maintenance and reliability data collection, analysis, and CBM-related work can easily be outsourced.

Succession Planning

- Succession planning is another key element of the workforce development process, with the goal to continuously identify and develop high-performing leaders for the future needs of the organization.
- "Five Keys to Successful Succession Planning" (Greengard)
    1. Identify key leadership criteria.
    2. Find future leaders and motivate them.
    3. Create a sense of responsibility within the organization.
    4. Align succession planning with the corporate culture.
    5. Measure results and reinforce desired behavior so that employees are prepared and trained for the jobs of tomorrow.

Develop a Leadership Culture

- Clarify Manager vs. Leader role
    - Manager manages (directs) based on yesterday's experience
    - Leader coach / inspire and helps to reach to your full potential and vision
    - They facilitate, guide, and encourage
- Establish and maintain a learning environment
    - Provide to personal and professional growth through education and training opportunities
- Be supportive to create a flexible work environment

Workforce Management Measures of Performance

- % training budget (could also be measured in $ and mhrs per employee)
- # certified professionals as of total workforce (could also be measured in %)
- # papers being presented or published at M&R-related conferences
- # people supporting/involved with industry-sponsored standardization or professional society groups

Summary

Big changes in today's workforce:

- People entering the workforce have a more diversified background and different attitudes about work.
- Today's workforce (Baby Boomers, Gen Xers, and Gen Yers) with its diversified backgrounds requires different sets of benefits as well as management styles to retain them.
- They want a life-work balance.
- And they want to be led, not managed — and certainly not micro-managed.

Summary (cont'd)

- Large % of managers have been trained in relatively autocratic and directive methods that are not welcomed by today's workforce.
- New employees entering the workforce, specifically in the M&R field, lack basic skill sets (including basic reading and math capabilities).
- Labor shortage will intensify in near future; therefore, finding qualified employees will be difficult and expensive.

Summary (cont'd)

- The organizational structure needs to be examined critically to ensure an optimal way for smooth operation
- Optimum organizational characteristics:
  - Supports productivity, accountability, and profitability
  - Workforce needs to be cross-trained to enable them to work productively together across departmental or functional lines
  - Adopts a more flexible workforce model in order to turn fixed cost into variable cost

Summary (cont'd)

Additional optimum characteristics:
- Established process for attracting, developing, and retaining key young talent, while minimizing loss of critical skills that older workers possess
- Long-term commitment to get workforce with appropriate knowledge and skill sets
- Shared services and outsourcing as part of a cost-effective solution
- Workforce aligned with overall business strategy
- Environment that continually challenges the workforce.

Chapter 10 Self-Assessment Questions

1. Who is Dr. W. Edwards Deming? What was his message?
2. How are Dr. Deming's principles related to the workforce?
3. Who were the quality gurus who revolutionized Japanese industry in the 1960s? What did they do?
4. Explain the employee life cycle. What is its importance?
5. What is the largest expense for an organization?
6. What does "Generation Gap" mean? How does it impact an organization?
7. What can we do to leverage different generations' employees to our advantage?
8. Why is the communication process important to us?
9. What happens in the communication process?
10. How should we determine training needs?

WORKFORCE MANAGEMENT

Ch 10 Comprehension Assessment Questions

1. What is one key to defining the difference between an ordinary company and a great organization?
2. Give Deming's 14 principles.
3. Why is it important to hire the best person in the first place?
4. What are some benefits of employees retiring after a long service time with an organization?
5. Choose one of the 4 basic generations of employees and describe it in detail, including their birth time period.
6. How do you motivate Gen Xers to maximize productivity?
7. What are the similarities between motivation factors for Gen Xers and Gen Yers?
8. What are some tools and strategies that can be used to communicate across generations?
9. What percentage of time do people in organizations typically spend in interpersonal situations to get work done?
10. Discuss the three basic listening modes.
11. What are the five elements of active listening?
12. What should a meeting facilitator do in order to ensure that the meeting is successful?
13. What are some ways of ensuring that time is not wasted in meetings?
14. What is job task analysis, and why is it important to a successful training program?
15. How can job analysis be used in the hiring/selection process?
16. What methods can be used to perform a job analysis?
17. What are the five categories of information collected by job analyst?
18. What are the benefits of an effective maintenance training program?
19. What are the basic elements that should be included within a training curriculum?
20. What are the different elements that an employee training database should include?
21. What are the benefits to employers for certifying their employees?
22. What are the four major classifications or groupings of certifications?
23. What are the primary reasons why people leave or change jobs?
24. What are the percentage of new entrants into the workforce today are women or minorities?
25. What is the definition of diversity as applied to the workplace?
26. What are the three types of organization structures that can be used to set up a maintenance reliability organization?
27. What are the maintenance and reliability functions that should be addressed when designing a maintenance and reliability organization structure?
28. Compare and contrast the advantages and disadvantages of a centralized vs. decentralized M&R organization structure.
29. Give a couple of examples of when outsourcing maintenance activities should be practiced.
30. What is succession planning?
31. What are the sudden actions that might take place that would preclude the need for an immediate succession plan implementation?
32. What is the ultimate goal of succession planning?
33. According to Greengard, what are the key steps for successful succession planning?
34. What are some different performance measures and benchmark data that can be used to evaluate workforce management and organization?

Research Questions

(Always cite your references when answering research questions.)

1. Discuss the history of Dr. W Edwards Deming. Contrast his experience with American manufacturers and improving organizations versus his experience with Japanese manufacturers improving their organizations.
2. Describe the metrics used within human resource processes and explain how they may negatively impact the overall effectiveness of the organization.
3. Evaluate the approach used by Disney in creating an employee culture that pushes employees to create happiness in their "guests."
4. Describe what at least two different companies are doing to inspire employees to perform to the best of their abilities.
5. Compare and contrast leaders and managers?
6. Discuss the negative aspects of a high-level of employee turnover within an organization.
7. Evaluate the advantages and disadvantages of having employees that are part of the "Silent" generation.
8. Evaluate the advantages and disadvantages of having employees that are part of the "Baby Boomer" generation.
9. Evaluate the advantages and disadvantages of having employees that are part of the "Gen X" generation.
10. Evaluate the advantages and disadvantages of having employees that are part of the "Gen Y" generation.
11. Discuss the advantages and disadvantages of providing work schedule flexibility rather than rigid workday start and stop times.
12. Compare different approaches that companies are using to improve their overall communication process.
13. Describe recent research related to the area of active listening.
14. Compare the advantages and disadvantages of having routine or regularly-scheduled team meetings.
15. Describe why developing people within an organization should be viewed as an investment.
16. Explain what a specific company is doing in order to ensure that their meetings are managed effectively.
17. Evaluate the literacy level of maintenance and operations personnel found in industry today compared to what is acceptable.
18. Discuss the advantages and disadvantages of organizations that have eliminated their apprenticeship programs.
19. Evaluate the advantages and disadvantages of incorporating certification and qualification as part of the overall maintenance training program.
20. Describe the certification process used by the Society of Maintenance and Reliability Professionals certifying organization.
21. Describe what a specific organization or organizations are doing to capture the knowledge of retiring baby boomer.
22. Explain in detail one or more company's programs used to manage diversity in their workplace.
23. Compare and contrast advantages and disadvantages of a centralized versus decentralized organization structure, including resources outside of this book.
24. Describe the advantages and disadvantages of outsourcing maintenance as a primary mode of completing maintenance activities.

25. Evaluate the advantages and disadvantages of establishing a small reliability and maintenance staff support group at the corporate level.
26. Discuss a specific example of how a major company has used succession planning to have a more effective organization.
27. Explain what companies are using to establish and maintain a constant learning environment within their organizations.
28. Describe the techniques being used in industry to improve the effectiveness of team meetings.
29. Evaluate how Deming's principles have been used to improve organizational effectiveness in industry.

Chapter 11: Maintenance Analysis and Improvement Tools

Every problem is an opportunity.
-- Kilchiro Toyoda, Founder of Toyota

Maintenance Analysis and Improvement Tools

11.1 Introduction
11.2 Key Terms and Definitions
11.3 Maintenance Root Cause Analysis Tools
11.4 Six Sigma and Quality Maintenance Tools
11.5 Lean Maintenance Tools
11.6 Other Analysis and Improvement Tools
11.7 Summary
11.8 Chapter Assessment
11.9 References and Suggested Reading

Chapter 11's Objective ... to understand:

- Why are analysis tools necessary?
- Types of analysis tools available
- What analysis tools should be used? When and where should they be used?
- Application of 6 sigma in a non-production environment
- What is meant by lean and VSM, and how they apply to maintenance

MAINTENANCE ANALYSIS AND IMPROVEMENT TOOLS

Introduction

- To remain competitive, organizations must continually:
  - Improve processes
  - Reduce costs
  - Cut waste
- ISO 9001 requires continuous improvement through data analysis and customer feedback for certification
- Industry surveys show most improvements do not reduce recurring problems because they address symptoms instead of the root causes.

Introduction (cont'd)

- O&M data should be analyzed to develop and implement improvements to assets/processes.
- Our assets/systems are getting very complex so that identification and diagnosis of problems is more rigorous and difficult.
- Our problem-solving ability has not improved.
  - Training too high level and philosophical
  - Not focused on analytical problem solving
  - Not taught to think logically/deductively
  - Lacking knowledge of what tools to use and how to apply them appropriately

Analysis and Improvement Tools

- Maintenance Root Cause Analysis (RCA) Tools
- Maintenance Six Sigma (6σ) and Quality Tools
- Lean Maintenance Tools
- Other Maintenance Analysis and Improvement Tools

CHAPTER 11

**What Is
Root Cause Analysis?**

**What Tools Can Be Used to
Perform This?**

Maintenance Root Cause Analysis

- If root causes of problems are not addressed in a timely fashion, they will be repeated.
- Root Cause Analysis (RCA)
  - Also known as Root Cause Failure Analysis (RCFA)
  - Structured methodology that leads to the discovery of the prime cause (or root cause)
  - Should address the root cause, not just symptoms

Maintenance Root Cause Analysis

- Assets, components, and processes can fail for a number of reasons but usually at a definite progression of actions and consequences that lead to the failure.
- RCA investigation traces the cause and effect trail from the failure back to the root cause.
  - Like a detective at work trying to solve a crime
  - Like the National Transportation Safety Board (NTSB) trying to piece together evidence following a plane crash
  - As with these, RCA is an iterative process

RCA Categories

- **Safety-Based RCA:** Performed to find causes of accidents related to occupational safety, health, and environment.
- **Product- or Production-Based RCA:** Performed to identify causes of poor quality, production, and other product-related problems in manufacturing.
- **Process-Based RCA:** Performed to identify causes of problems related to processes, including business systems
- **Asset-Based RCA:** Performed for failure analysis of assets/systems in the engineering and maintenance area

RCA Principles

- Aiming corrective measures at root causes is more effective than merely treating the symptoms of a problem.
- To be effective, RCA must be performed systematically, and conclusions must be backed up by evidence.
- There is usually more than one root cause for any given problem.

RCA Steps

- Define the problem (in maintenance, the failure).
- Collect data / evidence about issues that contributed to the problem.
- Identify possible causal factors.
- Develop solutions and recommendations.
- Implement the recommendations.
- Track the recommended solutions to ensure effectiveness.

RCA Tools and Techniques

- 5 Whys
- Cause and Effect Diagram (aka Fishbone or Ishikawa Diagram)
- Failure Modes and Effects Analysis (FMEA)
- Fault Tree Analysis

5 Whys Analysis

- A simple problem-solving technique that helps get to the root of the problem quickly.
  - Made popular in the 1970s by the Toyota Production System
  - Involves looking at any problem and asking "Why?" or "What caused this problem?"
  - Often, the answer to the first "Why?" will prompt another "Why?"
  - And so on multiple times (hence "5 Whys")
  - Quickly determines root cause of a problem
  - Easy to learn and apply

Example of 5 Whys Analysis

- Why is our customer (Ops Dept XYZ) unhappy?
  - We did not deliver our services on time.
- Why didn't we meet an agreed-upon schedule?
  - Job took much longer than we thought it would.
- Why did it take so much longer?
  - We underestimated the work requirements.
- Why did we underestimate the work?
  - We did not develop a work plan for this job.
- Why didn't we do planning for this job?
  - We were short on planner resources.

Clearly, we needed better planning.

MAINTENANCE ANALYSIS AND IMPROVEMENT TOOLS

Cause-and-Effects Analysis

- Analytical tool that provides a systematic way of investigating causes and effects.
  - Also known as a Fishbone or Ishikawa diagram
  - Named for its design (looks like the skeleton of a fish) and its designer (Dr. Kaoru Ishikawa, a Japanese quality control statistician)
  - It looks at the effects along with the causes that create or contribute to these effects

Cause-and-Effects Analysis Steps

1. Define the problem (head of the fish).
2. Brainstorm.
3. Identify all causes (the bones of the fish divided into major categories).
   - 6 Ms (Methods, Machines, Materials, Manpower, Measurement, and Mother Nature / Environment)
   - 4 Ps (Place, Procedure, People, and Policies)
   - 4 Ss (Surroundings, Suppliers, Systems, and Skills)
4. Select any causes not at the root of the problem.
5. Develop corrective action plan to eliminate or reduce the impact of the causes selected in previous step.

Example: Fishbone Diagram

Failure Modes and Effects Analysis

- A step-by-step methodology to identify all possible failures:
  - During the design of an asset (process)
  - Within a manufacturing or assembly process
  - In the operations or maintenance phase
  - When providing services
  - Also called Failure Modes, Effects, and Criticality Analysis (FMECA) or Potential Failure Modes and Effects Analysis
  - Purpose: To take actions to eliminate or reduce failures, starting with the highest priority ones

FMEA Questions

1. What are the components and the functions they provide?
2. What can go wrong?
3. What are the effects?
4. How bad are the effects?
5. What are the causes?
6. How often can they fail?
7. How can this be prevented?
8. Can this be detected?
9. What can be done; what design, process, or procedural changes can be made?

FMEA Process Steps

MAINTENANCE ANALYSIS AND IMPROVEMENT TOOLS

FMEA Benefits

- Early identification and elimination of potential asset/process failure modes
- Prioritization of asset /process deficiencies
- Documentation of risk and actions taken to reduce the risk
- Minimization of late changes and associated cost
- Improved asset (product) / process reliability and quality
- Reduction of Life Cycle Costs
- Catalyst for teamwork among design, operations, and maintenance

Fault Tree Analysis (FTA)

- A methodology that starts with the final failure and progressively traces each cause that led to the previous cause:
    - Each result clearly flows from its predecessor.
    - If evident a step is missing between causes, it is added in and its need is explained.
    - Once fault tree completed and checked for logical flow, investigating team determines changes needed to prevent or break the sequence of causes and consequences from occurring again.
    - Not necessary to prevent the first, or root cause, from happening – merely necessary to break the chain of events at any point so that final failure cannot occur.

Fault Tree Example

CHAPTER 11

What Do Six Sigma and Quality Have to Do with Maintenance?

Six Sigma and Quality

- Six Sigma is a quality improvement initiative, developed from Total Quality Management (TQM).
  - A disciplined, data-driven methodology for eliminating defects, driving toward six standard deviations in any process – from manufacturing to transactional and from product to service.
  - Major emphasis is its focus on reducing process variation to very low levels.
  - With many organizations, simply means a measure of quality striving for near perfection.

Six Sigma

Sigma Level	Defects per Million Opprt's - DPMO	Quality Level - %
1	691,462	30.85
2	308,537	69.15
3	66,807	93.32
4	6,210	99.38
5	233	99.977
6	3.4 (world class)	99.9999

Example - bag of potato or candy with
Mean weight = 200 g and standard deviation (σ) = 10 g
- 68% bag (190 – 210 grams) - 1 σ
- 95.5% bag (180 – 220 grams) - 2 σ
- 99.7% bag (170 – 230 grams) - 3 σ
- 0.3% bag (< 170 – > 230 grams)

*Based on Normal distribution

MAINTENANCE ANALYSIS AND IMPROVEMENT TOOLS

Deming Cycle vs. Six Sigma Process

Deming's PDCA Methodology:
- Plan
- Do
- Control (Check)
- Act

Six Sigma Methodology:
- D — Define
- M — Measure
- A — Analyze
- I — Improve
- C — Control

Pareto Analysis

- Commonly used across 6 Sigma and Quality
- Tool for visualizing the Pareto principle:
  - States that a small set of problems (the "vital few") affecting a common outcome tend to occur more frequently than the remainder.
  - 80% of the problems (failures) are caused by 20% of the items (assets / components)
  - <u>Pareto chart</u>: Bar graph arranging information to easily establish process improvement priorities and determine which subset of problems should be solved first (deserve the most attention)

Example: Pareto Chart

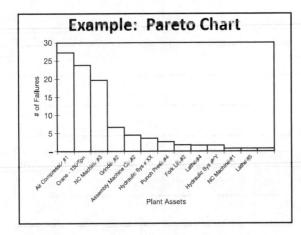

How Can Maintenance Ever Be "Lean?"

Lean Maintenance

- "Lean" is one of the new buzzwords in industry.
  - Lean literally means what it says ("Lean")
  - To be healthy, we need to get rid of fat (waste)
  - Similarly, in work environments, we need to be efficient and effective ("lean") to stay healthy and survive in today's competitive environment.
  - Kevin S. Smith (President, TPG Productivity, Inc.):
    - "Lean is a concept, a methodology, a way of working; it's any activity that reduces the waste inherent in any business process."
  - *Lean Thinking* (Womack & Jones) states that lean relates to customer-defined value.

"Muda" – The 7 "Wastes"

1. **Transportation**: Moving material and parts that are not actually required for the process
2. **Inventory**: Extra inventory in the system; all WIP (work-in-progress) and finished product not being processed
3. **Motion**: People or equipment moving or walking more than required to perform the work
4. **Waiting**: Waiting for the next step or something; time not being used effectively
5. **Overproduction**: Production ahead of demand or need
6. **Inappropriate Processing**: Extra steps in the process (use well-designed processes/assets)
7. **Defects**: Simplest form of waste because products not meeting specifications cause additional inspections and repairs

"Lean" Maintenance?

1. **Transportation**: Plan/provide materials and tools to reduce the # extra trips to the storeroom to hunt for right parts
2. **Inventory**: Eliminate/minimize extra inventory; keep only the right material/parts/tools in the storeroom
3. **Motion**: Minimize people movement (improved planning)
4. **Waiting**: Minimize waiting for the next step (another skilled person or part) by improved planning and scheduling
5. **Overproduction**: Develop optimized PMs, FMEA/RCM-based, perform root cause analysis to reduce failures, etc.
6. **Inappropriate Processing**: Use the right tools and fixtures to improve maintenance processes
7. **Defects**: Eliminate rework and poor workmanship; educate/train maintenance personnel appropriately

Value Stream Mapping (VSM)

Commonly used in Lean to understand and improve material/information flow within organizational processes:

- Processes have non-value-added (wasteful) activities to be identified and eliminated.
- VSM uses standard color-coding to assess value:
  - Green – Value-added activities
  - Yellow – Non-value-added activities required for regulatory (external/internal) requirements
  - Red – Non-value-added activities (waste)

The VSM Process

1. Identify the problem and set expectations.
2. Select the team.
3. Select the process to be mapped.
4. Collect data and produce current state map.
5. Critique / evaluate the current state.
6. Map the future state; identify waste and non-value-added items.
7. Create an action plan and implement.
8. Measure/evaluate results (readjust if needed).

What Are Other Analysis and Improvement Tools That Can Be Applied to Maintenance?

Theory of Constraints (TOC)

Improvement tool based around attacking process or system "bottlenecks":
- At any point in time, there is most often only one aspect of that system that is limiting its ability to achieve most of its goal (weakest link in the chain).
- For any significant improvement, its constraint must be identified and the whole system must be managed with this constraint in mind.
- If we want to increase throughput, we must identify and eliminate the constraint (or bottleneck).

The TOC Process

1. Communicate the goal of the process or organization (for example, "Make 100 units or more of X products per hour").
2. Identify the constraint (the factor within a process or system preventing organization from meeting goal).
3. Exploit the constraint; ensure constraint is dedicated to what it uniquely adds to the process, and not things that it should not do.
4. Elevate the constraint; if required, increase the capacity of the constraint with additional equipment or reducing set-up time.
5. If result of these steps moves constraint to another process or system, return to Step 1 and repeat.

Affinity Analysis

Tool to group ideas in a logical way to capture themes developed during brainstorming using the following steps:

1. Conduct a brainstorming meeting.
2. Record ideas and issues ("post-it notes" or cards).
3. Gather these into a single place (desk or wall) and sort into similar groups/patterns/themes based on the team's thoughts. Continue until all sorted and team is satisfied with groupings.
4. Label each group with a description of what it represents and place at the top of each group.
5. Capture and discuss the themes or groups and how they may relate.

Affinity Diagram Example

a) Random Ideas   b) Affinity Diagram

Barrier Analysis

Tool used particularly in process industries to trace energy flows and identify how and why barriers did not prevent damage.

- Can be used proactively (to design effective barriers and control measures) or reactively (to clarify which barriers failed and why).
- Barriers can be classified as:
  - Physical barriers
  - Natural barriers
  - Human action barriers
  - Administrative barriers

Summary

- Organizations must continually improve processes, reduce costs, and cut waste to remain competitive.
  - To make improvements, data must be analyzed
  - Tools and techniques that support corrective actions
- Approaches range from a simple checklist to sophisticated modeling software.
- Choices include 5 whys, cause–and-effects diagrams, Pareto analysis, root cause analysis, FMEA, Six Sigma, Lean, and Value Stream Mapping.
- Continuous improvement tools can optimize work processes, and improve results.
  - Provide a life-work balance.
  - Lead, don't manage — and certainly don't micro-manage.

Summary (cont'd)

- Major categories of analysis and improvement tools:
  - Maintenance Root Cause Analysis (RCA) Tools include 5 Whys, Cause-and-Effect Diagrams, FMEA, and FTA
  - Maintenance Six Sigma (6σ) and Quality Tools include PDCA, DMAIC, and Pareto Analysis
  - Lean Maintenance Tools (including VSM)
  - Other Maintenance Analysis and Improvement Tools include TOC, Affinity Diagram, and Barrier Analysis
- Applying continuous improvement tools can optimize work processes and help any organization to improve results, regardless of the size or type of business environment

Summary (cont'd)

Applying continuous improvement tools can optimize work processes and help any organization to improve results:

- Regardless of the size or type of business environment
- Including maintenance and reliability (M&R)
- Creativity and innovative thinking may be required to use many of these tools in M&R
- These tools can still be used to take your M&R organization to the optimum level for overall organizational success

MAINTENANCE ANALYSIS AND IMPROVEMENT TOOLS

Chapter 11 Self Assessment Questions

1. Into what categories can root cause analysis be classified?
2. What steps are needed to perform an RCA?
3. When should we use the fishbone tool?
4. How is a fishbone diagram constructed? Explain the steps that are needed.
5. What is the purpose of the FMEA?
6. What key steps / elements are needed to perform an FMEA?
7. What does 6 Sigma mean?
8. If a process is performing at 5 Sigma level, what percent of defective parts can be expected?
9. What is the difference between Deming's improvement cycle and DMAIC?
10. What is the primary objective of Pareto analysis?

Chapter 11 Other Comprehension Assessment Questions

1. What must organizations do to remain competitive?
2. What is a new requirement related to ISO 9001 that relates to analysis and improvement tools?
3. What are reasons why the ability for organizations to solve problems has not necessarily improved?
4. What are the major categories of maintenance analysis and improvement tools?
5. What is Root Cause Analysis (RCA)?
6. What is one negative impact of the root cause of the failure not being addressed in a timely fashion?
7. What percentage of problems is caused by processes versus personnel?
8. Were the general principles for root cause analysis?
9. What are the three basic types of root cause factors?
10. What are the different tools that can be used to perform root cause analysis?
11. What is a 5 Whys Analysis?
12. What is a Cause-and-Effects Analysis?
13. List at least 10 examples of "bones" that could be used for a Fishbone diagram.
14. What can make it difficult to use a Fishbone diagram to perform root cause analysis?
15. What is a Failure Modes and Effects Analysis (FMEA)?
16. When considering an FMEA, what are failure modes?
17. When considering an FMEA, what are effects?
18. What are the different categories of FMEAs?
19. How can FMEA be used to improve new design?
20. List the major FMEA-related published standards.
21. Define the term RPN.
22. What are the benefits of performing FMEAs?
23. What is a Fault Tree Analysis (FTA)?
24. What is the main difference between FTA and FMEA?
25. Given the following scenario, determine the appropriate root cause analysis tool that should be used. There is a single failure that would cause the entire plant to shutdown unexpectedly, and your plant manager has asked you to determine the appropriate risk mitigation strategy to minimize this potential.
26. Given the following scenario, determine the appropriate root cause analysis tool that should be used. You have responsibility for a specific group of equipment and need to develop an optimized maintenance strategy.
27. Given the following scenario, determine the appropriate root cause analysis tool that should be used. You are leading an investigation team to determine the root cause of why an environmental spill occurred.
28. Given the following scenario, determine the appropriate root cause analysis tool that should be used. You are leading a design team for a new product your company is launching and need to minimize the repair potential and cost of that item to capture market share for that product.
29. If your cell phone app manufacturer strives for 3 Sigma level of success in providing a working app initially to a customer, how many would be displeased every month if this app is accessed by 1,000,000 customers?

30. If your airline company were to strive for 4 Sigma level of success when it comes to not losing passenger bags, how many bags would be lost every day if they process on average 125,000 bags per day?
31. If your airline company were to strive for 6 Sigma level of flight success, how many unsuccessful flights would occur every year if they fly on average 1,000 trips per day?
32. What is meant by the Pareto principle?
33. Given the following failure data, which equipment should be further analyzed based on Pareto analysis?

Equipment	# Failures	Equipment	# Failures	Equipment	# Failures	Equipment	# Failures
1A	3	6F	39	11K	31	16P	38
2B	4	7G	7	12L	7	17Q	7
3C	26	8H	1	13M	40	18R	4
4D	6	9I	6	14N	2	19S	1
5E	4	10J	37	15O	33	20T	9

34. Describe what is meant by Lean with regard to a process organization. Lean literally means getting rid of the fat or waste within our work environment to be more efficient and effective, much like our own physical bodies (page 387).
35. What are three key Japanese words (and their corresponding English words) related to lean terminology?
36. List the seven categories of waste related to lean methodology.
37. Based on the seven categories of waste, give examples related to maintenance and reliability.
38. What is meant by Value Stream Mapping (VSM)?
39. What are the steps needed in order to carry out a VSM exercise?
40. Describe the different color coding categories related to VSM.
41. What is meant by the term Theory of Constraints?
42. What are the four primary process flow paths within the theory of constraints lexicon?
43. What are the key steps in implementing an effective theory of constraints approach?
44. What is meant by an affinity analysis or diagram?
45. Describe the process required to conduct an affinity analysis or diagramming exercise.
46. What is meant by the term barrier analysis?
47. What are the four major classifications of barriers in the context of barrier analysis?

Research Questions

(Always cite your references when answering research questions.)

1. Describe what specific organizations are doing related to ISO 9001 when implementing continuous improvement and data analysis within their standardized work processes.
2. Describe the Root Cause Analysis methodology used with at least one specific company.
3. Give specific examples of companies using 5 Whys analyses to solve a particular problem.
4. Give specific examples of companies using Cause-and-Effects analyses to solve a particular problem.
5. Give specific examples of companies using Failure Modes and Effects Analyses to solve a particular problem.
6. Give specific examples of companies using Fault Tree Analyses to solve a particular problem.
7. Choose one of the major standards associated with FMEAs and give a brief synopsis of that standard as well as at least one company's implementation of this standard.
8. Discuss the progression of at least one organization between Total Quality Management and 6 Sigma.
9. Explain what companies who were instrumental in beginning industry's journey toward 6 Sigma did to implement that methodology across their organization.
10. Describe at least one company's implementation of the PDCA cycle for continuous improvement.
11. Describe at least one company's use of the DMAIC process in implementing 6 Sigma.
12. Discuss at least one company's use of Pareto analysis as part of their overall maintenance and reliability improvement strategy.
13. Give a historical account of the development and first use of the Pareto principle by its founder.
14. Discuss details surrounding Dr. Joseph M. Juran's use of the Pareto principle in business applications.
15. Explain how "lean thinking" starts with the critical point of the customer defining value.
16. Compare and contrast the history of applying lean within manufacturing processes, specifically including elements of Henry Ford and the Toyota Production System.
17. Describe how specific companies are using the lean methodology within maintenance and reliability.
18. Give at least one company's use of Value Stream Mapping when improving their organization.
19. Give specific examples of companies using Theory of Constraints to solve a particular problem.
20. Give specific examples of companies using Affinity Analysis to solve a particular problem.
21. Give specific examples of companies using Barrier Analysis to solve a particular problem.
22. Describe at least one analysis or improvement tool not covered in this chapter that industry is using to solve a particular problem.

Chapter 12:
Current Trends and Practices

*"Progress is impossible
without change;
and those who cannot change
their minds,
cannot change anything "
— George Bernard Shaw"*

Current Trends and Practices

12.1 Introduction
12.2 Key Terms and Definitions
12.3 Energy Management, Sustainability, and the Green Initiative
12.4 Personnel, Facility, and Arc Flash Safety
12.5 Risk Management
12.6 Corrosion Control
12.7 Systems Engineering and Configuration Management
12.8 Standards and Standardization
12.9 Summary
12.10 Chapter Assessment
12.11 References and Suggested Reading

Chapter 12's Objective ... to understand:

- Why M&R leaders should worry about energy and sustainability or "being green"
- What M&R leaders should do about safety and, more specifically, arc flash safety
- How we go about managing the risks around us
- What is meant by corrosion control
- The effects of standards and standardization on your organization
- What exactly are systems engineering and configuration management

CURRENT TRENDS AND PRACTICES

Introduction

- Businesses must continually improve processes to remain competitive.
- Key means:
  - Awareness of current trends
  - Innovative best practices
- Further investigation is necessary:
  - Context in which a particular industry or business applies these trends and practices
  - How these trends and practices could be applied and implemented in your own organization

Introduction (cont'd)

Trends and practices that might apply to maintenance and reliability organizations:

- Energy and Sustainability
- Safety
- Risk Management
- Corrosion Control
- Systems Engineering/Configuration Management
- Standards

What about Sustainability?

How Do You Manage Energy?

What Is the Green Initiative?

What Is Sustainability?

- "Forms of progress that meet the needs of the present without compromising the ability of future generations to meet their needs" (World Commission on Environment and Development, 1987)
- Satisfaction of basic economic, social, and security needs now and in the future without undermining the natural resource base and environmental quality on which life depends
- Can foster policies that integrate environmental, economic, and social values in decision making

Pillars of Sustainability

Sustainability
- Economic
- Social
- Environmental

The Emergence of Sustainability

- All pillars should be coordinated and addressed to ensure the long-term viability of our community and the planet.
- The sustainability issue has emerged as a result of significant concerns about the unintended consequences of:
  - Rapid population growth
  - Economic growth
  - Increasing consumption of our natural resources

Our Role in Sustainability

Each person, business, and industry has a role and a responsibility to ensure their individual and collective actions support the sustainability of our community.
- Preserve our resources so that we can enjoy them in the future.
- Regenerate our resources at a rate that is equal to or faster than our consumption.

Sustainability and Business

- Concept of sustainability has evolved to reflect perspectives of both public and private sectors.
- Business goals of sustainability:
  - Increase long-term shareholder and social value
  - Decrease industry's use of materials
  - Reduce negative environmental impacts
- Favors an approach based on:
  - Capturing system dynamics
  - Building resilient and adaptive systems
  - Anticipating and managing variability and risk
  - Earning a profit

Good Sustainability Habits

1. Reduce what we buy and what we use.
2. Save on electricity-energy
3. Generate less waste
4. Reduce water resources
5. Minimize or eliminate need of hazardous material
6. Choose greener transportation anytime it is feasible

The Cost of Energy

- Energy supply chain begins with electricity, steam, natural gas, coal, and other fuels supplied to a manufacturing plant.
- Industrial energy systems account for roughly 80% of all energy used by the industry.
- Energy is a vital and often costly input to most production processes and value streams.
- Understanding/explicitly tracking energy costs can:
  - Show potential value of identifying and eliminating energy waste
  - Encourage energy conservation

Understanding Energy Usage

- A simple walkthrough is an excellent way to identify and fix energy wastes that are readily apparent.
- Two strategies for learning more about energy usage:
  - Measure the energy use of individual production and support processes.
  - Conduct an energy audit to understand how energy is used (and possibly wasted across facility).
    - Energy audits (or energy assessments) analyze/study energy end-uses and performance of a facility.
    - They can range in complexity and level of detail (simple facility walkthrough, review of utility bills, or comprehensive analysis of historical energy use and energy-efficiency investment options).

Identifying Energy Waste

- On average, more than 1/3 of energy we use is lost due to inefficient processes and waste.
  - Many examples exist across our facilities (machines left running, energy inefficient equipment and lighting, oversized equipment, air/vacuum leaks, heating and cooling, poor plant layout and flow, non-standard controls, etc).
  - As much as 50 % of this could be saved by improving the efficiency and reducing energy losses in these systems.
- Think of unnecessary energy usage as another "deadly waste" limiting business excellence.

Benefits of Energy Management

- Reduced operating and maintenance costs
- Reduced vulnerability to energy and fuel price increases
- Enhanced productivity
- Improved safety
- Improved employee morale and commitment
- Improved environmental quality, reduced greenhouse gas emissions, and remaining below air permitting emission thresholds
- Increased overall profit

Strategic Energy Management

- There are three steps involved in developing an energy planning and management roadmap appropriate to any organization:
  - Initial Assessment: Consider the opportunities, risks, and costs associated with strategic energy management.
  - Design Process: Understand organization's energy needs and identify best way to establish an energy mgmt plan.
  - Evaluate Opportunities: Identify/prioritize energy-related improvement opportunities (energy efficiency options, energy-supply options, energy-related products/services).
- New energy management standard has been issued (ISO 50001:2011) that establishes a framework to manage energy across multiple types of facilities.

The Green Initiative

- **Green Building Initiative:**
  - Initiative to reduce energy usage and environmental impact by challenging states to demonstrate leadership in energy efficiency and environmental responsibility in buildings
  - Requires states to reduce grid-based energy usage in its buildings 20% by 2015
- **Leadership in Energy & Environmental Design (LEED):**
  - Developed by the U.S. Green Building Council but internationally recognized green building certification with 3rd-party verification of energy/environmental strategies
  - Intended to provide concise framework for identifying and implementing practical and measurable green building design, construction, operations, and maintenance solutions.

CHAPTER 12

What Does Safety Have to Do with Maintenance and Reliability?

Relationship Between Safety and Reliability

- Safety and reliability historically considered two separate elements of the production operations system.
  - It has been observed that reliable plants are safe plants, and safe plants are reliable plants.
  - Furthermore, these plants are usually profitable plants, showing interrelationship between these.
  - Ron Moore, leading M&R expert, noted a strong inverse correlation (0.87) between OEE/Uptime and injury rate per 100 employees.

Creating a Safety Culture

- A Safety culture doesn't just stay at the organization or workplace – it goes home with us and becomes part of who we are.
- Developing a simple process model is a highly effective component of our overall strategy to improve safety in an organization, when focusing on:
  – Leadership
  – Personnel
  – Environment
  – Behavior
- Turning employees into Safety and Reliability Leaders creates an organizational "Culture of Caring."

CURRENT TRENDS AND PRACTICES

Creating a "Culture of Caring"

- Making safety a core value inspires a "Culture of Caring" within the entire workforce.
  - Many organizations (DuPont, Kimberly-Clark, Harley Davidson, General Mills, Milliken) have created an impressive safety culture, reducing injury and incident rates below 1/100 employees.
  - Jacobs Engineering's goal is 0 injuries, with a "Beyond Zero" initiative to create a "Culture of Caring," from the top throughout the organization.
  - "Everyone feels responsible for safety and pursues it on a daily basis; employees go beyond 'the call of duty' to identify unsafe conditions and behaviors, and intervene to correct them." [OSHA]

Challenge to Creating a "Culture of Caring"

DuPont's Rosanne Danner lists the most common reasons organizations fail to develop a safety culture:

- <u>Lack of Commitment from Leadership</u>: It's not profitability or safety (it's both), from the CEO through line management.
- <u>Inconsistency in How/Where Safety Applied</u>: Management must put in place the right procedures and consistently follow them.
- <u>Loss of Focus</u>: Instituting a safety culture not an overnight proposition; it will take time to become ingrained in an organization so we can't let quick results trick us into losing our long-term focus.

What Is Arc Flash Safety?

IEEE and the National Fire Protection Association (NFPA) have expressed a strong interest in the area of arc flash safety, developing the following definition:

- Arc flash is a strong electric current — and sometimes a full-blown explosion.
- It passes through air when insulation between energized conductors or between an energized conductor and ground is no longer sufficient to contain the voltage between them.
- This creates a "short cut" that allows electricity to race from conductor-to-conductor, to the extreme detriment of any worker standing nearby (hence the relationship to safety).

Dangers of Arc Flash

Arc flash resembles a lightning bolt-like charge, emitting heat to reach temperatures of 35,000°F in 1/1000 of a second.
- Massive energy released instantly vaporizes the metal conductors involved, blasting molten metal and expanding plasma outward with extreme force.
- This release can cause a fire, substantial damage to equipment, or even worse injury to personnel.
- Anyone exposed to the blast or heat without sufficient personal protective equipment (PPE) would be severely, and oftentimes fatally, injured.

Causes of Arc Flash

- Arcs can be initiated by a variety of causes:
  - Workers incorrectly think the equipment is de-energized and begin to work on it energized.
  - Workers drop or improperly use tools or components on an energized system.
  - Dust, water, or other contamination accumulate and cause insulation breakdown.
  - Connections loosen, overheat, and reach thermal runaway and fail.
- Potential equipment where this can happen include:
  - Panel boards, switchboards, and fused disconnects
  - Motor control centers, starters, and drive cabinets
  - Metal clad switchgear and transformers
  - Basically any place electrical equipment can fail

Arc Flash Regulations

OSHA requires employers to protect workers from arc flash exposure -- 1910.132(d) & 1926.28(a)
- Employer should assess these hazards in the workplace; select, have, and use correct PPE; and document this assessment.
- OSHA considers the NFPA 70E standard a recognized industry practice.
- Electrical inspectors now enforcing new labeling requirements in 2008 National Electric Code.

Arc Flash Compliance

•OSHA Arc Flash Compliance involves a 6-point plan:
1. A facility must provide, and be able to demonstrate, a safety program with defined responsibilities.
2. Establish shock and flash protection boundaries.
3. Provide protective clothing (PC) and personal protective equipment (PPE) that meet ANSI standards.
4. Train workers on the hazards of arc flash.
5. Provide appropriate tools for safe working.
6. Keep warning labels on equipment.
   - Note that the labels are provided by the equipment owners, not the manufacturers.
   - 110.16 NFPA 70E 2009 requires labels with the available incident energy or required level of PPE.

Arc Flash Prevention

- Worker training and an effective safety program can significantly reduce arc flash exposure.
- Considering fault currents, arcing time, and distance with design and equipment configuration choices.
- And implementing a comprehensive PM program:
  – Using corrosion-resistant terminals and insulating exposed metal parts, if possible
  – Sealing all open areas of equipment to ensure rodents and birds cannot enter
  – Verifying that all relays and breakers are set and operate properly
  – Using CBM technologies such as infrared and ultrasound where possible

How Can Anyone Possibly Manage Risk?

What Is Risk?

- Risk is potential that a chosen action or activity will lead to a loss, an undesirable event, or a particular outcome.
  - Simply stated, risk is a future event that has some uncertainty of occurrence and negative consequence if it were to occur.
  - Risk = Combination of the probability of an event and its consequences (ISO/IEC Guide 73).
  - In all types of tasks we undertake, there is potential for events and consequences that constitute opportunities for benefits or threats to success – this is risk.

Risk Management

- A technique that considers both positive and negative aspects of risk.
  - Should be a central part of any organization's strategic management process, methodically addressing risks associated with their activities
  - Focusing on the identification and treatment of these risks with the objective of adding maximum sustainable value to all activities of the organization
  - Understanding potential upsides and downsides of all factors that can affect the organization
  - Increasing probability of success and reducing both probability of failure and uncertainty of achieving organization's overall objectives

Risk Assessment

Risk Index (Magnitude) = Consequence (Impact) of Risk event x Likelihood (Probability) of Occurrence

Risk Assessment Matrix

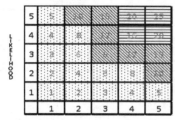

Risk Mitigation

- After risks have been identified and assessed, all techniques to manage the risk fall into one or more of these four major categories:
  - **A**voidance: Eliminate or don't do that activity
  - **C**ontrol: Optimize, mitigate, or reduce risk
  - **A**ccept: Accept and budget-plan
  - **T**ransfer: Risk share or outsource

What Is Meant by Corrosion Control?

What Is Corrosion?

Corrosion is the wearing-away of metals due to a chemical reaction.

- Naturally occurring phenomenon commonly defined as the deterioration of a substance, usually a metal, or its properties because of a reaction with its environment
- Disintegration of an engineered material into its constituent atoms due to chemical reactions with its surroundings
- Electrochemical oxidation of metals in reaction with an oxidant such as oxygen (e.g., formation of iron oxide aka rusting)

Corrosion's Impact

- Corrosion can cause dangerous and expensive damage
  - To automobiles, home appliances, water pipelines, bridges, buildings, and industrial plant infrastructure
  - So prevalent and takes so many forms that its occurrence and associated costs will never completely be eliminated
  - Total annual estimated direct cost of $276 billion (~ 3.1 % GDP) according to *Corrosion Costs and Prevention Strategies in the United States*"(U.S. Federal Highway Administration, 2002)
  - Estimated worldwide direct cost exceeding $1.8 trillion (~3-4% GDP of industrialized countries) according to other studies (China, Japan, United Kingdom, Venezuela)

Corrosion Control and Protection

- Four basic methods are needed for Corrosion Control and Protection:
  - Materials resistant to corrosion
  - Protective coatings
  - Cathodic protection
  - Corrosion inhibitors (modifying the operating environment)
- Can be achieved by incorporating the latest state-of-the-art technologies:
  - Original equipment design
  - In manufacturing
  - In maintenance, supply, and storage processes

What Do Systems Engineering (SE) and Configuration Management (CM) Have to Do with Maintenance?

SE vs. CM

- **Systems Engineering**: An interdisciplinary engineering management process that evolves and verifies an integrated, life-cycle balanced set of system solutions that satisfy customer needs
- **Configuration Management**: A discipline applying technical and administrative direction and surveillance to identify and document the functional and physical characteristics of an asset / system called a configuration item; control changes to those characteristics; and record and report changes

Key Primary Functions of SE

1. Design/Development
2. Build/Construction/Manufacture
3. Deployment/Fielding/Commissioning
4. Operations
5. Support/Maintenance
6. Disposal
7. Training
8. Verifications

The Role of CM

- Identify, examine, and select assets/systems, software, and documents.
- Periodically assesses program elements throughout the lifetime of the program.
- Record updates of all approved changes, accurately reflected in output documents (drawings, descriptions, specifications, and procedures).
- **Caution**: CM programs cost significant resources to implement. Evaluation of documents and records should be based on asset complexity, criticality, and a value-added assessment.

238 CHAPTER 12

How Do We Standardize Maintenance and Reliability?

What Is a Standard?

- "A prescribed set of rules, conditions, or requirements concerning definitions of terms;
- Classification of components;
- Specification of materials, performance, or operations;
- Delineation of procedures;
- Or measurement of quantity and quality in describing materials, products, systems, services, or practices"

As defined by National Standards Policy Advisory Committee

History of Standards

- Standards have evolved significantly:
  - Cylindrical stones in Egypt (7000 B.C.).
  - King Henry's length of his forearm (1120)
  - Boston's brick manufacturing (1689)
  - U.S. railroad system (Industrial Revolution)
  - The Baltimore fire and non-standard fire hydrants (1904)
  - Interchangeability of parts, components, and safety (20th century)
- American National Institute of Standards (ANSI) founded in 1918 via professional societies (ASME, IEEE, ASCE, ASTM, etc.) to support standards development.

Benefits of Standards

- Standards are a powerful tool for organizations of all sizes, while supporting innovation and increasing productivity, allowing organizations and whole industries to:
  - Implement and maintain best practices
  - Support safety of people and environment
  - Improve productivity — reduce cost
  - Attract and assure customers
  - Demonstrate market leadership
  - Create competitive advantage

Asset Management Standards*

- **ISO 55000**: Asset management – Overview, principles and terminology
- **ISO 55001**: Asset management — Management systems — Requirements
- **ISO 55002**: Asset management — Management systems — Guidelines for the application of ISO 55001

* Under development and scheduled to be released early 2014

Summary

- Being aware of current trends and innovative best practices allow businesses to continually improve processes to remain competitive:
  - Energy and Sustainability
  - Safety
  - Risk Management
  - Corrosion Control
  - SE/CM
  - Standards

Summary (cont'd)

Sustainability is important because it carries a reduction in energy usage by organizations:

- Sustainable development is defined as "forms of progress that meet the needs of the present without compromising the ability of future generations to meet their needs."
- Energy costs can have a significant impact on the financial performance of businesses.
- Energy reduction measures should be taken seriously by companies as the prices of energy continue to rise.

Summary (cont'd)

- Corrosion, a naturally occurring and quite prevalent phenomenon, can cause dangerous and expensive damage to everything and cannot be eliminated. Four basic methods of control and protection:
    - Materials resistant to corrosion
    - Protective coatings
    - Cathodic protection
    - Corrosion inhibitors
- Systems Engineering (SE) and Configuration Management (CM) should be considered not only for the products that are manufactured but also for the assets/systems maintained by an organization.

Summary (cont'd)

- Standards are rules or requirements determined by a consensus opinion of users and that prescribe the accepted and (theoretically) best criteria for a product, process, test, or procedure, with benefits to:
    - Safety
    - Quality
    - Reliability
    - Interchangeability of parts or systems
    - Consistency across international borders
- An Asset Management-specific standard should aid M&R organizations in establishing standards for their processes and becoming industry leaders.

Chapter 12 Self-Assessment Questions/Answers

1. **Define sustainability. Why is it important to organizations?** Sustainability is the ability to maintain a certain status or process in existing systems; in general refers to the property of being sustainable; capacity to endure. It is important to organizations because it increases long-term shareholder and social value while decreasing industry's use of materials and reducing negative impacts on the environment. (pages 404-405)

2. **What process improvement strategies can be used to reduce plant energy consumption?** In plant operations, several process improvement strategies (page 413) can be employed to reduce energy usage such as:
   - Total Productive Maintenance (TPM): Incorporate energy reduction best practices into day-to-day autonomous maintenance activities to ensure that equipment and processes run smoothly and efficiently.
   - Right-Sized Equipment: Replace oversized and inefficient equipment with smaller equipment tailored to the specific needs of manufacturing.
   - Plant Layout and Flow: Design or rearrange plant layout to improve product flow while also reducing energy usage and associated impacts.
   - Standard Work, Visual Controls, and Mistake-Proofing: Sustain and support and energy performance gains through standardized work procedures and visual signals that encourage energy conservation.

3. **What four major categories of equipment/systems use the majority of energy in the industry, as defined by DOE?** Steam; process heat; motors, pumps and fans; and compressed air. (pages 409-410)

4. **Generally, the electricity bill is broken down by what types of charges? What can be done to minimize the total electric energy cost?** The basic factors that determine an industrial power bill are:
   - Kilowatt hour consumption
   - Fuel charge adjustments
   - Kilowatt demand
   - Power factor penalty (in some cases)

   To minimize total electricity bill, we need to reduce electricity consumption, total demand (power-KW), and bad power factor.

5. **Define the major categories of risk to which a product (asset) or project may be exposed.** (pages 433-434)
   - Safety risk
   - Performance risk
   - Cost risk
   - Schedule risk
   - Technology risk
   - Product data access and protection risk

6. **Why is configuration management important? Discuss its application in the maintenance – asset management area.** Configuration management (CM), a component of SE, is a critical discipline in delivering products that meet customer requirements and that are built according to approved design documentation. In addition, it tracks and keeps updated system documentation which includes drawings, manuals, operations/maintenance procedures, training, etc. CM is the methodology of effectively managing the life cycle of assets and products in the plant. It prohibits any change of the asset's form, fit, and function without a thorough, logical process that considers the impact proposed changes have on life cycle cost. (page 438)

7. **What strategies are used to reduce the impact of arc flash hazards?** (page 425)
    - Provide a demonstrated safety program with defined responsibilities.
    - Establish shock and flash protection boundaries.
    - Provide protective clothing (PC) and personal protective equipment (PPE) that meet ANSI standards.
    - Train workers on the hazards of arc flash.
    - Provide appropriate tools for safe working.
    - Apply warning labels on equipment.
8. **Why do we use standards? How can they be classified?** We use standards to achieve a level of safety, quality, and consistency in the products and processes that affect our lives. Standards can be classified in two categories: Specifications (codes) and Process improvement (management). (page 444)
9. **What is the intent of the ISO 55000 family of standards?** The overall purpose of these three International Standards is to provide a cohesive set of information in the field of Asset Management Systems that will: (page 445-446)
    - Enable users of the standards to understand the benefits, key concepts, and principles of asset, asset management, and asset management systems.
    - Harmonize the terminology being used in this field.
    - Enable users to know and understand the minimum requirements of an effective management system to manage their assets.
    - Provide a means for such management systems to be assessed (either by the users themselves, or by external parties).
    - Provide guidance on how to implement the minimum requirements.

APPENDIX

1.7 Self-Assessment Questions / Answers

1. Define a best practice. What are the barriers to implementing best practices?

   An idea that asserts there is a technique, method, or process more effective at delivering a desired outcome than any other technique, method, or process. (p. 2)
   Also, 1. Knowledge about current best practices, 2. Motivation to make changes for the adoption of best practices, 3. Knowledge and skills required to make changes to adopt best practices. (p. 3)

2. What are keys factors that impact the performance of plant machinery?

   1. Inherent reliability, 2. Operating environment, 3. Maintenance plan. (p. 4-5)

3. Why has reliability become a comparative advantage in today's environment?

   Reliability initiatives in an organization reduces operating costs by reducing failures and optimizing maintenance costs, thereby providing competitive advantages. (p. 5)

4. Identify five key performance measures in the area of maintenance and reliability. Elaborate on each element of these measures. What are typical "world class benchmark numbers" for these performance measures?

   1. Maintenance Cost as a Percentage of Replacement Asset Value (RAV); Best Practice Benchmark = 2-9%, Typical World Class = 2.0-3.5%
   2. Maintenance Material Cost as a Percentage of Replacement Asset Value (RAV); Best Practice Benchmark = 1-4%, Typical World Class = 0.25-1.25%
   3. Schedule Compliance; Best Practice Benchmark = 40-90%, Typical World Class = >90%
   4. Percentage Planned Work; Best Practice Benchmark = 30-90%, Typical World Class = >85%
   5. Production Breakdown Losses; Best Practice Benchmark = 2-12%, Typical World Class = 1-2%
   6. Parts Stock-Out Rate; Best Practice Benchmark = 2-10%, Typical World Class = 1-2%
   (pg. 6)

5. Define what makes a benchmark "World Class." Discuss, using specific examples.

World Class, also called *Best in Class,* are the organizations that have achieved higher performance, better than others in their class or quartile.
The Maintenance Cost as Percent of RAV benchmark for best in class (1st / 2nd Quartile) organizations may vary between 2 –5 %, but world class organizations are usually in < 2.5 %, still keeping same or better quality/reliability standards. (p. 6, 304)

6. Define planned work and identify its benefits. What is a typical world class benchmark number?

The work that has gone through a formal planning process to identify work that needs to be accomplished with work sequence, materials needed, labor skills, tools, safety requirements etc. The objective is to perform the work efficiently and effectively. The world class benchmark is > 85 %. (p. 6-7)

2.9 Self-Assessment Questions / Answers

1. What are the key attributes of a leader?

    Visionary, energizing people, communicator, competent, charismatic. (p. 27,28)

2. Why is vision important? Also define vision statement.

    A vision defines what an organization wants to become. Corporate success depends on the vision articulated by the organizational leaders and management.

    A vision statement is a short, succinct, and inspiring declaration of what the organization intends to become or to achieve at some point in the future; refers to the category of intentions that are broad, all-inclusive, and forward-thinking; the image that a business must have of its goals and before it sets out to reach them; the aspirations for the future, without specifying the means that will be used to achieve those desired ends. (p. 29, 30)

3. Define MBWA. Why is it considered one of the key leadership practices?

    Management By Wondering Around. Simply it means: Get out of the office and circulate among the troops. (p. 26)

4. Define an organizational culture

    Culture refers to an organization's values, beliefs, and behaviors; it is the beliefs and values which define how people interpret experiences and behave both individually and in groups. (p. 22)

5. What are the key benefits of having a mission statement?

    They help organizations focus their strategy by defining some boundaries within which to operate; they define the dimensions along which an organization's performance is measured and judged; they suggest standards for individual ethical behavior.

    It is an organization's vision translated into written form; the leaders view of the direction and purpose of the organization; a vital element in any attempt to motivate employees and to give them a sense of priorities. It's simply, what we are planning to do and what will be accomplished. (p. 28,32)

6. Why are mission and vision statements important for an organization?

    An organizations needs to establish strategic framework to be successful. The foundation of this framework is vision and mission statements. They establish strategic vision and mission that state where organization needs to go and how they plan to get there. (p 28-29)

7. Define reliability culture.

When all, working together as a team, take proactive actions to eliminate or minimize failures, take appropriate action when a failure situation happens, and total cost of ownership is one of the key criteria to reduce over-all cost. The results are: lower lost time – emergency repairs, improved safety and availability and with lower operating costs. (p. 38-42)

8. Why is change management an important part of creating the right reliability culture?

Creating a reliability culture involves proactive thinking and involvement of all in the organization. This is a major paradigm shift which requires change management strategy to accomplish. (p36-39)

9. Define the role of a change agent. Who is best qualified to perform this role?

Change agent is someone who has clout, conviction, charisma, resourcefulness etc. to make things happen and keep the stakeholders involved in implementing change. Generally, this role is filled by a senior management person who has respect and trust of the people. (p. 38)

10. State the difference between a manager and leader

A manager is an individual who is in charge of a certain group of tasks, employees or certain subset of an organization. Generally, a manager is concerned about today and makes decision based on past (experience) whereas leader is a person or thing who may not hold a management position but is able to exercise a high degree of influence over others. A leader is more concerned with the future and doing the right things.

SELF-ASSESSMENT QUESTIONS AND ANSWERS

3.9 Chapter Self-Assessment Questions / Answers

1. **Define maintenance and its role.**
   Maintenance is the work of keeping assets in proper condition, and keeping them from losing partial or full functional capabilities; preserve /protect asset functions; keeping assets – equipment in good working condition so that the assets can be used to their full capacity when needed. (pg. 50)

2. **What are the different categories of maintenance work?**

   PM, CBM/PdM, Proactive, and Corrective. (pg. 55-59)

3. **What can equipment operators do to support maintenance?**

   Operators are the first line of defense against unplanned asset downtime. They are with the assets all the time and can use their knowledge and skills to predict and prevent breakdowns and other losses. (pg. 60)

4. **Why would an organization support operators getting involved in maintenance?**

   Operators are with the assets all the time. They can detect any abnormalities and symptoms at an early stage and get them corrected before they can turn into major failures. It's more cost effective practice. (pg. 61)

5. **Why would an organization need to have a CMMS? What is the difference between a CMMS and an EAM?**

   A CMMS (Computerized Maintenance Management System) is an essential tool for maintenance organizations for documenting and processing work associated with specific assets. The new CMMS are known as Enterprise Asset Management (EAM) and they usually have much broader functional capabilities and can easily be integrated with other business systems. (pg. 63)

6. **List five maintenance metrics and discuss why they are important.**

   1) Maintenance cost as % of ERV
   2) Unscheduled CM (reactive) cost as % of total maintenance cost
   3) PM cost as % of total maintenance cost
   4) Planned work as % of total work
   5) % Overtime

   They help to improve maintenance effectiveness.

7. **Name five PdM technologies and discuss how they can help reduce maintenance costs.**

   Some of the technologies are Vibration Analysis, Infrared Thermography, Acoustic/Ultrasonic Sound Level Measurements, Oil Analysis, Electrical Testing, and Partial Discharge.

All of these technologies can help to identify potential failures earlier and reduce/ optimize maintenance costs. (pg. 55-57)

8. Define Proactive Maintenance

All work tasks completed to avoid failures or to identify defects that could lead to failures. Any maintenance that is performed proactively, such as CBM/PdM, PM, or those things identified by CBM or PM; only those actions that are identified by CBM/PdM or PM; anything on the maintenance schedule that has been identified in advance, and is planned and scheduled. (pg. 58-59)

9. What are the benefits of a structured maintenance program?

Reducing production downtime (the result of fewer asset failures); increasing life expectancy of assets, thereby eliminating premature replacement of machinery and assets; reducing overtime costs and providing more economical use of maintenance personnel due to working on a scheduled basis, instead of an unscheduled basis, to repair failures; reducing cost of repairs by reducing secondary failures; reducing product rejects, rework, and scrap due to better overall asset condition; identifying assets with excessive maintenance costs, indicating the need for corrective maintenance, operator training, or replacement of obsolete assets; and improving safety and quality conditions. (pg. 54-55)

10. How can a CMMS/EAM system help improves maintenance productivity?

It can help to improve a maintenance department's efficiency and effectiveness and ultimately, to get more out of assets and reduced costs by streamlining critical workflows, work identification, work task planning, scheduling, reporting etc. (pg. 63)

SELF-ASSESSMENT QUESTIONS AND ANSWERS

4.10 Chapter Self-Assessment Questions /Answers

1. **Draw a workflow chart to show work from a request to completion.**

   See Fig. 4.3 and compare your chart

2. **Explain each role as shown in the workflow chart from Q 4.1.**

   Key roles are:
   - Work requester: Who makes the request of needs?
   - Coordinator – Asset/resource: Who established the priority and arranges resources?
   - Planner: Who plans the job/tasks?
   - Scheduler: Who schedules the job/tasks and ensures all the resources are available?
   - Work execution Supervisor: Who reviews work from an execution perspective view and ensures quality of work performed?
   - Work performer: Who performs the work?
   - Configuration Manager/Engineer: Who ensures that configuration changes are documented properly?
   - Materials Coordinator: Who arranges/expedites material?

   (Pg. 88-90)

3. **What is the purpose of a job priority system?**

   It allows ranking of work orders/tasks to be accomplished in order of importance. (pg. 96)

4. **Why do we need to manage maintenance backlog? What is a good benchmark?**

   Backlog indicates how much work is ready to be performed. It help to keep maintenance resources utilized effectively. A 4–6 weeks of backlog of work is a good benchmark. (pg. 113)

5. **What are the symptoms of ineffective planning?**

   Maintenance people standing around waiting on parts; high rework; poor work performance; stock-out in the storeroom; planners being used to expedite parts; maintenance personnel arriving at the jobsite and waiting for the asset/system to be shut down; frequent trips to storeroom by maintenance personnel; production downtime always more than estimated. (page 102)

6. **Should planners help schedulers or craft supervisors during an emergency? If yes, explain.**

   A planner is someone who plans the work/tasks for the future. Ideally, the planner should not be involved with day-to-day work such as supporting emergencies. (page 105)

7. **Who are key players in scheduling process? Explain their roles.**

Scheduling process determines when and who is going to do the work-job. It is a process by which resources are allocated to a specific job based on operational requirements and resources availability. Maintenance, operations and maybe the material-storeroom play the key role in this process. (pg.105–107)

8. **What are the key differences between planning and scheduling processes?**

    Planning: What and how
    Scheduling: When and who
    (pg. 99-107, 114)

9. **Discuss work types and the benefits of work classifications.**

    - PM: Preventive Maintenance
        - Time based (TBM)
        - Run based (RBM)
        - Condition based (CBM)
        - Operator based (OBM)
    - CM: Corrective Maintenance
        - CM – Routine – resulting from PM
        - CM – Major scheduled repairs
        - CM – Reactive/breakdowns

    Work classification help us to optimize our resources.

10. **What are the key differences between capital projects and turnarounds?**

    Scope: Capital projects are well defined and static drawings are available whereas turnarounds are loosely defined and dynamic with changes as inspections are made with regard to scope.

    Planning and scheduling: Capital projects can be planned and scheduled well in advance whereas with turnarounds, planning and scheduling cannot be finalized until scope is approved.

    Safety permits: Capital projects have fixed safety permits on a weekly or monthly basis whereas turnarounds require shift and daily basis due to scope fluctuations.

    Manpower staffing: Capital projects are fixed and usually don't change very much whereas turnarounds are variable, changing a lot during execution due to scope fluctuations.

    Schedule updates: Capital projects are updated on a weekly or bimonthly basis whereas turnarounds are updated on a shift or daily basis. (page 108)

SELF-ASSESSMENT QUESTIONS AND ANSWERS

Chapter 5 Self-Assessment Questions / Answers

1. **Inventories in a plant are generally classified into what categories?** Finished goods, work in process, raw materials, and maintenance and operating items such as spare parts and operating supplies including consumables. (page 122)

2. **What is meant by ABC classification as it is related to inventory?** Inventory is classified based on item value and usage rate to distinguish between the trivial many and the vital few, typically into the three categories of ABC. (page 124-125)

3. **Discuss how the cost of inventory can be optimized.** Inventory cost can be optimized by an analysis technique known as Economic Order Quantity (EOQ) analysis that determines the optimum quantity at specific time intervals to minimize overall inventory cost while still meeting customer needs. (page 138)

4. **How will you organize a storeroom? Discuss the key features of a small storeroom you have been asked to design.** The physical layout of the storeroom should be planned for efficient material flow. Some key features are: the storeroom should be separated from the main plant operations either by walls or the secured page to discourage pilferage of tools and extensive items; the parts-materials area should be sized and equipped appropriately for these items to be stored, keeping heavy parts low, close to, or on the floor; parts that are slow movers should be stored in the back of the storeroom while fast movers are stored in the front for easy and fast access; oil supplies should be kept away from the main storage area; each storage location and parts storage bin properly labeled; storage area should be free of clutter and debris to ensure personnel can move around to access parts easily and should have sufficient lighting; and storeroom should receive returns with sufficient space available to handle returns. (pages 128-129)

5. **Why is inventory accuracy important? What will you do to improve it?** Inventory accuracy is important for the following reasons: if the part cannot be found in the location indicated in CMMS records, the repair may not be completed on time; an out-of-stock condition can occur because parts are not ordered on time since actual quantity may be less than indicated in CMMS records; ordering when not required due to inaccurate counts in CMMS records may result in unnecessary inventory; and inaccurate inventory records may cause personnel to lose confidence, which may encourage hiding items. One can improve inventory accuracy with the following: recording all parts-material received against a purchase order in inventory system/CMMS; recording additional specific information related to parts (such as manufacturer's number, serial number, lot size, cost, and shelf life); recording all issued parts-material on a work order accurately along with the employee name, number, equipment, and projects; and returning all parts-material not used after a repair or PM back to the correct location and recording into the system. (pages 135-136)

6. **What are the key factors used in calculating EOQ?** Demand/usage in units per year, ordering cost per order, and inventory carrying cost per unit per year. (page 138)

7. **Explain the benefits of using RFID technology to label stock items–material.** Reduced inventory control and provisioning costs; accurate configuration control and repair history; part installation and removal time tracking; accurate and efficient parts tracking; reduced parts receiving costs; elimination of data entry errors; improved parts traceability; and reduced risk of unapproved parts. (page 145)

8. **Explain inventory turnover ratio. What are the benefits of tracking this ratio?** Inventory turnover ratio is the number of times an organization's investment in inventory is recouped during an accounting period, calculated by the value of issued inventory divided by the average inventory value. The benefits of tracking this ratio (amongst other measures) on a regular basis is to evaluate the improvements that have been made or to set up improvement goals. (page 147)

9. **Identify three key performance measures that can be used to manage MRO storerooms effectively.** Any of the following 10 measures could be identified: % Inactive Inventory; % Classification; Inventory Variance (Inaccuracy); Service Level; % Inventory Cost to Plant Value; Inventory Shrinkage Rate; % Vendor Managed Inventory (VMI); Inventory Growth Rate; % Stock Outs; and Inventory Turnover Ratio. (page 145-147)

10. **What is meant by *shelf life*? What should be done to improve it?** Shelf life is the amount of time that a specific stock item can be used. It can be improved or sustained by periodic/preventive maintenance (PM), proper storage, and identification of shelf life characteristics in CMMS as part of a shelf-life management program. (page 135)

Chapter 6 Self-Assessment Questions / Answers

1. **Define reliability and maintainability.** Reliability is the probability that an asset or item will perform its intended functions for a specified period of time under stated conditions. It is usually expressed as a percentage and measured by Mean Time Between Failures (MTBF). Maintainability is the ease and speed with which a maintenance activity can be carried out on an asset. It is a function of equipment design and usually measured by Mean Time to Repair (MTTR). (pages 156-157)

2. **What's the difference between maintenance and maintainability?** Maintenance is the act of maintaining or the actual work of keeping an asset in proper operating condition. Maintainability is the ease and speed with which a maintenance activity can be carried out on an asset. Therefore, one (maintainability) is a measure of the ease at which the other (maintenance) is performed. (page 153)

3. **If an asset is operating at 70% reliability, what do we need to do to get 90% reliability? Assume assets will be required to operate for 100 hours.** Reliability is a function of MTBF and operating time; therefore, in order to achieve 90% reliability when the current value is 70% reliability, one must increase the current MTBF value.

    MTBF = -(100)(1/(ln(0.70))) = 280 hours/failure
    MTBF = -(100)(1/(ln(0.90))) = 949 hours/failure

    Therefore, in order to get 90% reliability, this asset's MTBF must be increased from 280 to 949 hours/failure. (example shown on page 164)

4. **If an asset has a failure rate of 0.001 failures/hour, what would be the reliability for 100 hours of operation?**

    Reliability = $e^{-\lambda t}$ = $e^{-(0.001)(100)}$ = 0.9048 = 90.48%   (example shown on page 163)

SELF-ASSESSMENT QUESTIONS AND ANSWERS

5. **What would be the availability of an asset if its failure rate is 0.0001 failures/hour and average repair time is 10 hours?**

    MTBF = $1/\lambda$ = 1/(0.0001) = 10,000 hours/failure

    MTTR = repair time = 10 hours/failure

    Availability = MTBF/(MTBF + MTTR) = 10,000/(10,000+10) = 0.999 = 99.9%

    (example shown on page 166)

6. **What would be the availability of a plant system if it is up for 100 hours and down for 10 hours?**

    Uptime = 100 hours   Downtime = 10 hours

    Availability = Uptime/(Uptime + Downtime) = 100/(100 + 10) = 0.9091 = 90.91%

    (example shown on page 166)

7. **If an asset's MTBF is 1000 hours and MTTR is 10 hours, what would be its availability and reliability for 100 hours of operation?**

    Availability = MTBF/(MTBF + MTTR) = 1,000/(1,000 + 10) = 0.9901 = 99.01%

    Reliability = $e^{-t/MTBF}$ = $e^{-(100)/(1000)}$ = 0.9048 = 90.48%

    (example shown on pages 166-167)

8. **Define availability. What strategies can be used to improve it?** Availability is the probability that an asset is capable of performing its intended function satisfactorily, when needed, in a stated environment; it is a function of both reliability and maintainability (page 156). Availability can be improved with strategies to improve either reliability, maintainability, or both; trade-off studies should be performed to evaluate the cost effectiveness of strategies to increase MTBF (reliability) or decrease MTTR (maintainability) (page 160). Because reliability is designed-in, it can only be improved by redesigning it or replacing it with better components (page 155). Availability (through maintainability attributes) can be improved by repairing or replacing bad components before they fail and by implementing a good reliability-based PM plan. (page 155)

9. **What is the impact of O&M cost on the total life cycle cost of an asset?** On average, Operations & Maintenance (O&M) costs are about 80% of the total life cycle cost of the asset. (page 177)

10. **What approaches could we apply during the design phase of an asset to improve its reliability?** The following approaches should be taken during the design phase to improve an asset's reliability (pages 178-182):
    - Design so that the asset can be easily operated and maintained with minimum operations and maintenance needs.
    - Involve asset owners and operators in developing requirements as well as in reviewing the final design.
    - Focus attention on reliability requirements and specifications.
    - Select and configure proper components.
    - Design for fault tolerances, to fail safely, with early warning of failure to the user, with built-in diagnostic system to identify fault location, and to eliminate all or at least the critical failure modes cost-effectively.

Chapter 7 Self-Assessment Questions / Answers

1. **Explain operator-driven reliability. Why is the operator's involvement important in maintenance?** Under the operator-driven reliability concept, operators perform basic maintenance activities beyond their classic operator duties. Operator involvement encourages production to interact with maintenance and other departments as a team to reduce the number of failures, thus improving plant-wide asset reliability. (page 193)

2. **Define TPM. What are TPM's various elements (the pillars of TPM)?** Total Productive Maintenance (TPM), a maintenance strategy that originated in Japan, emphasizes operations and maintenance cooperation; its goals include zero defects, zero accidents, zero breakdowns, and an effective workplace design to reduce overall operations and maintenance costs. (page 191) There are 8 different elements (pillars) of TPM: autonomous maintenance; focused improvement; planned maintenance; quality maintenance; training and development; design and early equipment management; office improvement; and safety, health, and environment. (pages 197 through 201)

3. **How do we implement TPM?** There are 10 steps required to implement TPM: announce TPM; launch a formal education program; create an organizational support structure; establish basic TPM policies and quantifiable goals; outline a detailed master deployment plan; kickoff TPM; improve the effectiveness of each piece of equipment; conduct training to improve operations and maintenance skills; develop an early equipment management program; and continuous improvement. (pages 201-202)

4. **Define OEE. How do we measure it?** Overall Equipment Effectiveness (OEE) is a measure of equipment or process effectiveness based on actual availability, performance, and quality of product or output. It is measured by multiplying these three factors and expressing this calculation as a percentage. (page 191)

5. **What is the difference between OEE and TEEP?** The only difference between OEE and TEEP is a 4th factor of utilization that is included in the TEEP calculation but not in the OEE calculation. (page 191)

6. **Explain 5S. What benefits do we derive from implementing 5S?** 5S is a structured program to achieve organization-wide standardization in the workplace. The benefits derived from implementing 5S are a safer, more efficient, and productive operation. (page 190)

7. **What is the difference between 5S plus or 6S (and traditional 5S)?** The additional element that is the difference between 5S plus/6S and traditional 5S is safety. (page 208)

8. **Explain what is meant by the visual workplace.** A visual workplace uses visual displays to relay information to employees and guide their actions by setting up a workplace with signs, labels, color-coded markings, etc., so that anyone unfamiliar with the assets or process can readily identify what is going on, understand the process, and know both what is being done correctly and what is out of place. (page 192)

9. **Explain Muda, Mura, and Muri.** Muda is defined as waste, Mura is defined as inconsistencies, and Muri is defined as unreasonableness. (page 192)

10. **What are the benefits of standardizing?** Standardizing enables easier identification of anomalies and abnormalities and correction of such anomalies/abnormalities immediately. It also avoids reverting to old work habits and makes the process simpler and easier. (page 208)

Chapter 8 Self-Assessment Questions / Answers

1. **What is RCM? How did it get its start? Tell a little about RCM's history.** Reliability-centered maintenance (RCM) is a systematic and structured process to develop an efficient and effective maintenance plan for an asset to minimize the probability of failures and ensure safety and mission compliance. (page 223) RCM was developed in the commercial aviation industry to optimize maintenance and operations activities. (page 220)
   The history of RCM started with Boeing's 747 jumbo jets. The FAA required all its owner operators to get their PM programs approved by them in order to get certified for operation. Because of this very extensive maintenance program requirement, the airlines thought they may not be able to operate the 747 jumbo jet in a profitable manner. Therefore, they undertook a complete re-evaluation of their PM strategy. The result was a new approach employing a decision-tree process that prioritized critical PM tasks. This new approach was labeled RCM and was expected to be applied to all major military systems. Over the many years since then, RCM has matured significantly, but industry has yet to fully embrace this RCM methodology. (pages 226-227)
2. **Which Standards Development Organization (SDO) developed RCM Standard JA1011?** The Society of Automotive Engineers (SAE). (page 228)
3. **Describe the 4 principles of RCM. What is the key objective of RCM analysis?** Principal 1: The primary objective of RCM is to preserve system function. Principal 2: Identify failure modes that can defeat the functions. Principle 3: Prioritize function needs (failure modes). Principal 4: Select applicable and effective tasks. The key objective of RCM analysis is to maintain the inherent reliability of system function. (pages 227-228)
4. **During what phases of asset development do we get the maximum benefit of a RCM analysis? Why?** During the design and development phases of the assets to eliminate or mitigate the effects of failure modes. (page 226)
5. **Describe the 9-step RCM analysis process.** 1. System selection and information collection; 2. System boundary definition; 3. System description and functional block diagram; 4. System functions and functional failures; 5. Failure Modes and Effects Analysis (FMEA); 6. Logic (decision) Tree Analysis (LTA); 7. Selection of maintenance tasks; 8. Task packaging and implementation; 9. Making the program a living one through continuous improvement. (page 229-230)
6. **Which type of failure mode is not evident to the asset operator?** Hidden failure modes. (page 239)
7. **What are the benefits of RCM?** Safety, improved reliability, overall reduced cost, improved documentation, replaced parts and equipment, and increased efficiency and productivity. (pages 243-244)
8. **What is meant by CBM and PdM? What methods are used to perform these?** Condition-Based Maintenance (CBM) or Predictive Maintenance (PdM) is maintenance based on the actual condition (health) of assets obtained from in place, noninvasive measurements and tests. (page 223) There are multiple methods (or technologies) used to perform CBM/PdM: flow rates, temperature, pressure, electrical data, vibration, ultrasonic testing (contact and airborne), lubricant analysis, and infrared thermography. (pages 247-248)

9. **What is the difference between diagnostic and prognostic analysis?** Diagnosis isolates the cause of the problem while prognosis develops a corrective action plan based on its condition and remaining life. (page 245)

10. **What is velocity analysis? With which CBM technology is it associated? What does a peak at twice rotational speed indicate?** Velocity analysis is one piece of vibration technology with its data displayed in a graph that shows vibration velocity (expressed in inches/second) on the vertical axis and frequency on the horizontal axis. (page 250) If this type of graph shows a peak at twice rotational speed, then a bent shaft is likely a problem. (pages 250-251)

SELF-ASSESSMENT QUESTIONS AND ANSWERS   257

Chapter 9 Self-Assessment Questions / Answers

1. **Why do we need a performance measurement system? What are the benefits of such a system?** Performance measurement systems are needed in order to make the improvements needed for staying in business in a competitive market place. (page 284) The benefits of a performance measurement system is its ability to provide a factual basis in the following areas: strategic feedback to show the present status of the organization from various perspectives, diagnostic feedback of various processes to guide improvements on a continuous basis, trends in performance over time as the metrics are tracked, and feedback around the measurement methods themselves in order to track the correct metrics. (pages 284-285)

2. **What are the benefits of benchmarking?** The benefits of benchmarking are the opportunity to gain a strategic, operational, and financial advantage and the ability to blend continuous improvement initiatives and breakthrough improvements into a single change management system. (page 298)

3. **Explain what is meant by a "World Class" benchmark.** A benchmark that would be ranked by customers and industry experts to be among the best of the best and exemplary performance achieved independent of industry, function, or location. (page 288)

4. **What are the key attributes of a metric?** Metrics should encourage the right behavior, should be difficult to manipulate to "look good," and should not require a lot of effort to measure. (page 288)

5. **Explain leading and lagging metrics.** Leading indicators are forward-looking and help manage the performance of an asset, system, or process whereas lagging indicators tell how well we have already managed. (page 291)

6. **What types of metrics show results?** Lagging metrics (indicators) are results that occur after the fact. (page 291)

7. **Explain the Balanced Scorecard model.** The Balanced Scorecard Model is a strategic management approach developed in the early 1990s by Dr. Robert Kaplan of Harvard Business School and Dr. David Norton. It identifies four perspectives from which to view a process or an organization: Learning and Growth, Business Process, Customer, and Financial. A drawing of Figure 9.3 could also be used to answer this question. (pages 292-293)

8. **Explain the different types of benchmarking. What are the benefits of external benchmarking?** There are two types of benchmarking activities: internal and external. The benefits of external benchmarking are the ability to compare similar business processes or compare to the best in industry. (page 299)

9. **Discuss data collection and quality issues. How can we improve data quality?** Data can be easy or difficult to collect, and emphasis must be placed on data quality. (page 284) A key challenge with performance measurement systems is data collection and availability of quality data in a timely basis. (page 297) Any of the following questions could be used to address how to improve data quality, but it is probably best to identify the number of these questions for the student to answer. (pages 297-298)
    - Do the metrics make sense? Are they objectively measurable?
    - Are they accepted by and meaningful to the customer?
    - Have those who are responsible for the performance being measured been fully involved in the development of this metric?
    - Does the metric focus on effectiveness and/or efficiency of the system being measured?
    - Do they tell how well goals and objectives are being met?
    - Are they simple, understandable, logical, and repeatable?

- Are the metrics challenging but at the same time attainable?
- Can the results be trended? Does the trend give useful management information?
- Can data be collected economically?
- Are they available in a timely manner?
- Are they sensitive? (Does any small change in the process get reflected in the metric?)
- How do they compare with existing metrics?
- Do they form a complete set—a balanced scorecard (i.e., adequately covering the areas of learning and growth, internal business process, financial, and customer satisfaction)?
- Do they reinforce the desired behavior—today and in the long haul?
- Are the metrics current (living) and changeable? (Do they change as the business changes?)

10. **List five metrics that can be used to measure overall plant level performance of maintenance activities. Discuss the reason for your selection.** Any of these could be selected with your subjective evaluation of the reasons for the specific selections: Maintenance Cost as % RAV, Maintenance Cost per Unit Output, Return on Net Assets, Percentage Overtime, Training Hours/Person, Safety Performance (OSHA Injuries per 200K hours), Availability, Downtime as % Total Scheduled (Operating) Hours, or Overall Equipment Effectiveness. (page 304)

Ch 10 Self-Assessment Questions / Answers

1. **Who is Dr. W. Edwards Deming? What was his message?** Dr. W. Edwards Deming was a world-renowned expert in the field of quality. He taught us many quality management principles which any organization can use to improve its effectiveness. These principles can be applied to any organization, small to large, and to any industry (services, manufacturing, etc.). Dr. Deming is best known for reminding management that most problems are systemic and that it is management's responsibility to improve the systems so that workers (management and non-management) can do their jobs more effectively. Deming argued that higher quality leads to higher productivity, which in turn, leads to long-term competitive strength. (pages 310-312)

2. **How are Dr. Deming's principles related to the workforce?** Most of Dr. Deming's principles relate to the workforce, including management and their role in acquiring, preparing, and educating the workforce as well as improving the processes to get productivity gains. (pages 310-312)

3. **Who were the quality gurus who revolutionized Japanese industry in the 1960s? What did they do?** Dr. W. Edwards Deming, Dr. Joseph Juran, Philip Crosby, Armand V. Feigenbaum, etc. are known as the "Quality Gurus." They made a significant impact on the world through their contributions to improving not only businesses, but all organizations including state and national governments, military organizations, educational institutions, healthcare organizations, and many other establishments and organizations. They helped organizations by teaching quality principles, specifically to Japanese in the 1960s and 1970s.

4. **Explain the employee life cycle. What is its importance?** It is the entire useful life of an employee, from hiring to retiring. The employee life cycle consists of four steps: hire, inspire, admire, and retire. During this time, a person learns new skills and becomes a productive employee until retirement. (pages 314-315)

5. **What is the largest expense for an organization?** Employees are one of an organization's largest expenses these days. Unlike other major capital costs such as buildings, machinery, and technology, human capital is highly volatile. Managers are placed into key positions to reduce that volatility by reducing the overall life cycle cost of employees in the organization. (page 315)

6. **What does "Generation Gap" mean? How does it impact an organization?** People born in the last 90 years or so, since 1920 to now, can be grouped into four groups or "generations" (Silent, Baby Boomers, Generation X, and Generation Y), with each generation having their own attributes and expectations. The differences between the generations create many challenges in the workplace. These challenges, which can be both negative and positive, often relate to variations in perspective and goals as a result of generational differences. This area gets further complicated because of the age differences between managers and employees. Organizations can't assume that people of varying ages will understand each other or have the same perspectives and goals. In order to be successful, there is a need to understand and value the generational differences and perspectives and turn those negatives into positives. (pages 417-418)

7. **What can we do to leverage different generations' employees to our advantage?** In order for an organization to be truly successful, all co-existing generations in the workplace need to understand and value each other, even when their perspectives and goals are different. Organizations need to look beyond the clash of the generations for ways to leverage multi-generational perspectives to their benefit. We need to learn a few tools and strategies for communicating across generations. This would help us to tap the best that each generation as well as each individual brings to the workplace. (page 322)

8. **Why is the communication process important to us?** People in organizations typically spend over 75% of their time in interpersonal situations to get work done. It is no surprise to find that poor communications are at the root of a large number of organizational problems. Effective communication is an essential component of organizational success whether it is at the interpersonal, inter-group, intra-group, organizational, or external level. Effective communication is all about conveying messages to other people clearly and unambiguously. It's also about receiving information that others send to us, with as little distortion as possible. (page 323)

9. **What happens in the communication process?** Effective communication exists between two people when the receiver interprets and understands the sender's message in the same way the sender intended it. By successfully getting the message across, we convey our thoughts and ideas effectively. When not successful, the thoughts and ideas that we actually send do not necessarily reflect what we think, causing a communications breakdown and creating roadblocks that stand in the way of our and the organization's success—both personally and professionally. The communication process has several stages (Sender (source), Encoding, Channel (media), Decoding, Receiver, Feedback), with each stage having the potential for misunderstanding and confusion. To be effective communicators and to get our point across without misunderstanding and confusion, our goal should be to reduce the frequency of problems at each stage of this process with clear, concise, accurate, well-planned communications. (pages 323-324)

10. **How should we determine training needs?** A well-developed training program should be based on a job task analysis and skills assessment. The training must be focused to produce results as quickly as possible and must also meet an organization's long-term goals. (pages 330-334)

Chapter 11 Self-Assessment Questions /Answers

1. **Into what categories can root cause analysis be classified?** Root cause analysis is not a single, defined methodology; most of these can be classified into four, very broadly defined categories based on their field of application: safety-based, production-based, process-based, and asset failure-based. (page 359)
2. **What steps are needed to perform an RCA?**
   The Six Steps in Performing an RCA (page 360):
   - Define the problem — the failure.
   - Collect data / evidence about issues that contributed to the problem.
   - Identify possible causal factors.
   - Develop solutions and recommendations.
   - Implement the recommendations.
   - Track the recommended solutions to ensure effectiveness.
3. **When should we use the fishbone tool?** It is helpful to use the fishbone diagram in the following (page 366):
   - To stimulate thinking during a brainstorming session
   - When there are many possible causes for a problem
   - To evaluate all the possible reasons when a process is beginning to have difficulties, problems, or breakdowns
   - To investigate why an asset or process is not performing properly or producing the desired results
   - To analyze and find the root causes of a complicated problem and to understand relationships between potential causes; to dissect problems into smaller pieces
4. **How is a fishbone diagram constructed? Explain the steps that are needed.**
   The following five steps are essential when constructing a fish-bone diagram (page 367):
   1. Define the problem.
   2. Brainstorm.
   3. Identify all causes.
   4. Select any causes that may be at the root of the problem.
   5. Develop corrective action plan to eliminate or reduce the impact of the causes selected in Step 4.

   In general, the following steps are taken to draw the fishbone diagram (page 368-369):
   1. List the problem/issue to be investigated in the "head of the fish."
   2. Label each "bone" of the "fish."
      a. The team may use one of the categories suggested above, combine them in any manner, or make up others as needed. The categories are to help organize the ideas.
      b. Use an idea-generating technique (e.g., brainstorming) to identify the factors within each category that may be affecting the problem or effect being studied. The team should ask, "What are the issue and its cause and effect?"
      c. Repeat this procedure with each factor under the category to produce sub-factors. Continue asking, "Why is this happening?" and put additional segments under each factor and subsequently under each sub-factor.

d. Continue until you no longer get useful information when you ask, "Why is that happening?"

e. Analyze the results of the fishbone after team members agree that an adequate amount of detail has been provided under each major category. For example, look for those items that appear in more than one category. These become the most likely causes.

f. For those items identified as the most likely causes, the team should reach consensus on their priority. The first item should be listed the most probable cause.

5. **What is the purpose of the FMEA?** The purpose of an FMEA is to identify all possible failures during the design of an asset (product) — in a manufacturing or assembly process, in the operations and maintenance phase, or in providing services. (page 370)

6. **What key steps / elements are needed to perform an FMEA?**
Figure 11.4 (page 372) depicts the sequence in which a FMEA is performed with the typical sequence of steps answers the following set of questions (page 373):
   1. What are the components and the functions they provide?
   2. What can go wrong?
   3. What are the effects?
   4. How bad are the effects?
   5. What are the causes?
   6. How often can they fail?
   7. How can this be prevented?
   8. Can this be detected?
   9. What can be done; what design, process, or procedural changes can be made?

7. **What does 6 Sigma mean?** 6 Sigma is a statistical concept that measures a process in terms of defects; achieving 6 Sigma means that the process is delivering only 3.4 defects per million opportunities. (page 382)

8. **If a process is performing at 5 Sigma level, what percent of defective parts can be expected?** 233 defects per million opportunities translates to 0.0233% defective parts based on 99.977% quality level (page 383 includes table, but 5 Sigma level is incorrect in the book – corrected here).

9. **What is the difference between Deming's improvement cycle and DMAIC?** The classic Deming's PDCA (Plan, Do, Control and Act) cycle is used to plan improvements, implement and test them, evaluate if they worked, and then standardize if they did. However, for problem solving in Six Sigma, the PDCA cycle has been modified slightly to have a five-phase methodology called DMAIC (Define, Measure, Analyze, Improve, Control). The DMAIC method involves completing the necessary steps in a sequence. Skipping a phase or jumping around will not produce the desired results. It is a structured process to solve problems with proper implementation and follow-up. (page 384)

10. **What is the primary objective of Pareto analysis?** The primary objective of a Pareto chart is to determine which subset of problems should be solved first, or which problems deserve the most attention. (page 385)

Chapter 12 Self-Assessment Questions/Answers

1. **Define sustainability. Why is it important to organizations?** Sustainability is the ability to maintain a certain status or process in existing systems; in general refers to the property of being sustainable; capacity to endure. It is important to organizations because it increases long-term shareholder and social value while decreasing industry's use of materials and reducing negative impacts on the environment. (pages 404-405)

2. **What process improvement strategies can be used to reduce plant energy consumption?** In plant operations, several process improvement strategies (page 413) can be employed to reduce energy usage such as:
   - Total Productive Maintenance (TPM): Incorporate energy reduction best practices into day-to-day autonomous maintenance activities to ensure that equipment and processes run smoothly and efficiently.
   - Right-Sized Equipment: Replace oversized and inefficient equipment with smaller equipment tailored to the specific needs of manufacturing.
   - Plant Layout and Flow: Design or rearrange plant layout to improve product flow while also reducing energy usage and associated impacts.
   - Standard Work, Visual Controls, and Mistake-Proofing: Sustain and support and energy performance gains through standardized work procedures and visual signals that encourage energy conservation.

3. **What four major categories of equipment/systems use the majority of energy in the industry, as defined by DOE?** Steam; process heat; motors, pumps and fans; and compressed air. (pages 409-410)

4. **Generally, the electricity bill is broken down by what types of charges? What can be done to minimize the total electric energy cost?** The basic factors that determine an industrial power bill are:
   - Kilowatt hour consumption
   - Fuel charge adjustments
   - Kilowatt demand
   - Power factor penalty (in some cases)

   To minimize total electricity bill, we need to reduce electricity consumption, total demand (power-KW), and bad power factor.

5. **Define the major categories of risk to which a product (asset) or project may be exposed.** (pages 433-434)
   - Safety risk
   - Performance risk
   - Cost risk
   - Schedule risk
   - Technology risk
   - Product data access and protection risk

6. **Why is configuration management important? Discuss its application in the maintenance – asset management area.** Configuration management (CM), a component of SE, is a critical discipline in delivering products that meet customer requirements and that are built according to approved design documentation. In addition, it tracks and keeps updated system documentation which includes drawings, manuals, operations/maintenance procedures, training, etc. CM is the methodology of effectively managing the life cycle of assets and products in the plant. It prohibits any change of the asset's form, fit, and function without a thorough, logical process that considers the impact proposed changes have on life cycle cost. (page 438)

7. **What strategies are used to reduce the impact of arc flash hazards?** (page 425)
    - Provide a demonstrated safety program with defined responsibilities.
    - Establish shock and flash protection boundaries.
    - Provide protective clothing (PC) and personal protective equipment (PPE) that meet ANSI standards.
    - Train workers on the hazards of arc flash.
    - Provide appropriate tools for safe working.
    - Apply warning labels on equipment.
8. **Why do we use standards? How can they be classified?** We use standards to achieve a level of safety, quality, and consistency in the products and processes that affect our lives. Standards can be classified in two categories: Specifications (codes) and Process improvement (management). (page 444)
9. **What is the intent of the ISO 55000 family of standards?** The overall purpose of these three International Standards is to provide a cohesive set of information in the field of Asset Management Systems that will: (page 445-446)
    - Enable users of the standards to understand the benefits, key concepts, and principles of asset, asset management, and asset management systems.
    - Harmonize the terminology being used in this field.
    - Enable users to know and understand the minimum requirements of an effective management system to manage their assets.
    - Provide a means for such management systems to be assessed (either by the users themselves, or by external parties).
    - Provide guidance on how to implement the minimum requirements.